Un-Natural Discourse in the Age of Anthropogenic Landscapes

Un-Natural Discourse in the Age of Anthropogenic Landscapes

How We Imagine Wildlife

Barbara K. Jones

LEXINGTON BOOKS
Lanham • Boulder • New York • London

Lexington Books
Bloomsbury Publishing Inc, 1385 Broadway, New York, NY 10018, USA
Bloomsbury Publishing Plc, 50 Bedford Square, London, WC1B 3DP, UK
Bloomsbury Publishing Ireland, 29 Earlsfort Terrace, Dublin 2, D02 AY28, Ireland
www.rowman.com

British Library Cataloguing in Publication information available

Library of Congress Cataloging-in-Publication Data

Names: Jones, Barbara K., author.
Title: Un-natural discourse in the age of anthropogenic landscapes : how we imagine wildlife / Barbara Jones.
Description: Lanham, Maryland : Lexington Books, [2025] | Includes bibliographical references and index. | Summary: "This book argues how by relying on unnatural discourse to relate to the natural world, coexistence becomes much more difficult to achieve"— Provided by publisher.
Identifiers: LCCN 2024052879 (print) | LCCN 2024052880 (ebook) | ISBN 9781666914801 (cloth) | ISBN 9781666914818 (epub)
Subjects: LCSH: Nature—Effect of human beings on. | Animals and civilization. | Landscape protection. | Human ecology.
Classification: LCC GF75 .J64 2025 (print) | LCC GF75 (ebook) | DDC 304.2—dc23/eng/20241209
LC record available at https://lccn.loc.gov/2024052879
LC ebook record available at https://lccn.loc.gov/2024052880

For product safety related questions contact productsafety@bloomsbury.com.

∞™ The paper used in this publication meets the minimum requirements of American National Standard for Information Sciences—Permanence of Paper for Printed Library Materials, ANSI/NISO Z39.48-1992.

For my family and all the wildlife on our shared Earth

Contents

Acknowledgments

This book could not have been written without the help and support of my family and the many wonderful people I have met during the research and writing process. My research took me—and at times my husband, Tommy, and our daughter, Kelsey—around the United States, to the Arctic regions of Norway, and then across Europe. Along the way, I attended elk and wolf management meetings in the Intermountain West, caribou hearings in Canada, and talks about bison, wolves, grizzly bears, and beavers. I went kayaking and hiking with informants, attended conferences, and had conversations with people from New York City to Svalbard about their thoughts on wildlife and wildlife management. To initially learn about caribou and reindeer, I spoke with Greg Finstad who provided me with stories about the important role wild caribou and semi-domesticated reindeer have played in Alaska's ecology, history, and culture. Gregg Servheen helped me make sense of the gray ghost metaphor and gave me critical information about woodland caribou biology. Mark McCollough, John Pastor, and Matthew LaRoche shared their insights into woodland caribou reintroduction efforts in the lower forty-eight United States, while Mike Schrage, Noah Greenwald, Lydia Allen, David Argument, Gregg Servheen, and Cory DeStein explained to me the social and biological controversies woodland caribou recovery faces in Canada and the United States.

Bison as livestock is a metaphor that needed to be fleshed out, and the voices of Shana Drimal, Chamois Andersen, Donald Forgione, and Rick Lamplugh helped me appreciate how the language of livestock stripped the bison of their wildness and locked them behind fences in most places around the United States. Roger Auster and Malou Anderson-Ramirez and her family helped me see how coexistence with beavers, wolves, and grizzly bears is truly possible if you navigate the gray space in which people live. To Malou Anderson-Ramirez, the way questions are framed to affected stakeholders like ranchers or hunters can result in better outcomes for species many despise. Mark Heinz spoke to me about a recent wolf "whacking" episode that occurred in Wyoming in 2024. That event and our conversation helped me tie cultural heritage to language and to a way of seeing the wolf as a species long disdained in Wyoming, while a public conversation by Doug Smith and Chris Servheen about the gray wolf and grizzly bear, respectively, showed me the logic of talking about those species together.

The beaver story was an amazing one. Its transition from a critical species to Native people in North America, to a commodity and eventually a pest to European settlers, to its status today as an ecosystem engineer is an important example of the role imagining species plays in shaping how we see them. Mike Callahan, Suzanne Fouty, Aaron Hall, and Mark Vieira helped me see how circular the beaver's story has been; today, more of us are seeing the beaver's presence on the land in a positive way. The beaver-as-a-pest metaphor has lost some of its stickiness and has been replaced by one that sees the beaver as a water storer and habitat builder. That could not have happened if the beaver had not been reimagined.

And finally, this book could not have been written without the willingness of the many people who shared their thoughts and knowledge with me. My gratitude to them and the many unnamed in this book is deeply felt. That gratitude is also for my family and friends who helped me not only think through my ideas but also to read versions of my evolving manuscript. Many times, my husband and our daughter traveled with me, read and offered suggestions, and were patient as I interviewed, participated with, and observed people's interactions with nature. My sister, Cindy, helped with the nitty-gritty paperwork, while my friend Laura read and advised me on several of the manuscript's chapters. Through the efforts of many kind and helpful readers, I was able to make better sense of the cultural and biological issues my chosen species face.. Even though it does take a village to write a book, any errors made are mine alone.

Information from these sources was used in writing *Un-Natural Discourse in the Age of Anthropogenic Landscapes: How We Imagine Wildlife*:

Jones, Barbara K. "Unifying America One Bison at a Time." *Anthropology News* website, March 26, 2021.

Jones, Barbara K. *Wild Capital: Nature's Economic and Ecological Wealth*. Gainesville. University of Florida Press, 2019.

Introduction

When Wendell Berry wrote in his 1968 essay called "A Native Hill," "We have lived by the assumption that what was good for us would be good for the world,"[1] he spoke to our hubris that allowed us to assume what was good for us was good for the world. He concluded that we were wrong in those assumptions, yet fifty-five years later, it is obvious that in many places our hubris and arrogance persist. For us to have contributed to the sixth extinction alone should be extremely problematic to us all, but the failure to admit on a global scale the need to use natural resources in a sustainable way suggests an ignorance of natural processes and the limits we have in managing them. Our willingness to politicize wildlife, water use, or climate change indicates the natural world that sustains us is framed in a somewhat capricious cultural context rather than an evidence-based scientific one. That cultural context resonates because it is maintained through enduring metaphors and stories closely tied to an individual's identity. Those opposed to accepting the language of coexistence as a path to sharing the Earth with other species rely on ideas rooted in the past and a worldview embedded in a cultural heritage that sees nature as subservient to us, while those in favor of coexistence as the ideal look through a lens of mutualism or interdependent relationships,[2] where what is good for the world will be good for us.

Although behaviors associated with our cultural heritage or traditions of living are important for maintaining connections to our past, it is not a fixed concept. Culture changes, and with that change, the stories and metaphors forged by cultural heritage must also shift. A worldview that tolerated early conservationists killing what could have been the last American bison or fur seal makes no sense in today's world; just as Adolph Leopold who was hired to help exterminate gray wolves in the first half of the twentieth century had to change his thinking to eventually become one of the most important voices of a new and evolving environmental ethic. Romanticizing the past so individuals can maintain their cultural heritage intact sanitizes that past into something ideal and pure, yet when predator control in the nineteenth and early twentieth centuries relied on killing animals with strychnine, so they suffered an agonizing end, or baited them with fishhooks that tore through their guts, so they bled to death, the past was a much meaner and less forgiving place. In that setting, human-wildlife stories that recognized this brutality as cruel and unhealthy couldn't compete with the pithy and powerful metaphors that had helped normalize a savage domination over nature as the ideal. An attachment to this relationship made shifting the stories and reframing the metaphors to fit an emerging new environmental ethic a challenging one.

A landscape without large predators misses what makes the land wild, just as prairies without bison or boreal forests without woodland caribou diminish those ecosystems. Hunters claim to enjoy hunting because it takes them into nature, yet many fail to hunt according to the principles of fair chase and legitimate use, while ranchers and loggers who pride themselves on knowing the land and the species that share it, simultaneously encourage predator, bison, and pest removal laws. This unwillingness to coexist on the land with all manner of species is offset by the logger whose living is earned by cutting down virgin boreal forests in Canada critical to caribou survival but who will use that same logging equipment to help build caribou maternity pens in British Columbia, or a rancher who would prefer no bison, wolves, or grizzly bears near his livestock yet will remove the lowest strand of a barbed wire fence so pronghorns can migrate safely across his land. This willingness to accommodate the needs of wildlife reminds us that although cultural heritage can rely on an imagined past to define a person's contemporary values, it is neither fixed nor natural. With stakeholder conversations that can lead to reimagined possibilities for human-wildlife interactions, cultural heritage can be an important tool for teaching about the past rather than an excuse for maintaining those same behaviors today.

CULTURAL HERITAGE AS THE REASON

Cultural heritage or tradition has created many of the stories and metaphors people rely upon to assign wildlife its place in our world. Assigning words like pests, vermin, or varmint, or metaphors like "a good wolf is a dead wolf" or "beaver are water rats" are ways to make a species appear unpopular, strange, or even evil. By equating beavers with rats, the native beaver that might have caused few problems gets categorized with invasive rats, while calling a prairie dog a varmint diminishes its lived existence to one that encourages and tolerates its violent death. Stories and metaphors like these that are rooted in past human-wildlife relationships have made shared attitudes about coexistence and mutualism difficult to achieve.

Despite that, it is important to recognize that the value of cultural heritage to a person's sense of identity can't be discounted. It ties us to a shared past and sense of community, just as it tethers us to earlier generations. However, what it doesn't and shouldn't do is allow a heritage rooted in another time and place to justify a relationship with nature today. For many people, cultural heritage is a way to explain their animosity toward certain wildlife, their doubts about our role in climate change, or their intolerance for changing regulations or rules about land use. Yet those same behaviors that made sense in their family or community's nineteenth- or early twentieth-century past might no longer apply. When European settlers came with their Old-World ideas and met the New World with its unfamiliar demands, they had to change and adapt or fail in this new landscape. Today's descendants of many of those early settlers find themselves faced with change again as a domination-over-nature relationship that worked for hundreds of years and was believed to be fixed and natural is shifting to one based on conservation and greater efforts at coexistence. By clinging to the stories and metaphors passed down through family or community lore as if those stories remained a fixed reality has undermined what it means to coexist in a shared landscape today.

Through pest or vermin narratives, human-defined boundaries that keep humans separate from wildlife have become a societal norm. Anthropogenic landscapes that contain highly fragmented landscape features with rigid boundaries such as fences, houses, roads, dams, and agricultural plots, as well as suburban and urban settings, reaffirm that separation. Fences keep some things in but also work to manage unwanted animals seeking entry from eating our garden plants, digging under our sheds, or killing birds at our feeders. Another kind of boundary is even more challenging for wildlife because it relies on imaginary borders that exist in the ways we assign property ownership. A prairie dog or wolf living in a national park does not recognize the finite nature of this relatively safe habitat, and in their quest for food, mates, or better habitat might leave a national park's invisible boundary and enter a space where killing a wolf because a rancher "hates predators" or shooting a prairie dog because it is a "pest" are legitimate human behaviors in that less tolerant space. This adversarial relationship with predator or pest species is maintained by metaphors and stories that make taking these animal's lives rather than trying to coexist with them appear to be natural and appropriate. Through competing domination over mutualism narratives that generate different human-wildlife relationships, it becomes possible for wolves to be seen simultaneously as the wilderness ideal in some protected landscapes and vermin in private or less protected public spaces.

For our protected lands, whether they are national, state, or local parks, it is essential to see them not as isolated islands of protected land but as conservation centers of intersecting regions within a broader landscape. To achieve this, we need to think beyond those boundaries and ensure that the lands surrounding the parks do not act as brick walls to wildlife but offer uninterrupted stretches of protected land that provide habitat for wildlife, maintain biodiversity, and support critical ecosystem functions. Increasing wildlife access to these intersecting regions requires us to rethink the way we see wild species in both our imaginations and on the land.

Early tales by fiction and nonfiction nature writers encouraged the idea of boundaries as something both people and wildlife recognized. Ernest Thompson Seton, an author of works Theodore Roosevelt called "nature fakers" to identify them as "sham natural historians" who produced fake natural histories, wrote in the late 1890s about how the trash dump at the Fountain Hotel in Yellowstone National Park (YNP) offered him a way to observe bears feeding. Seton's observations at the bear-feeding grounds shaped his depiction of YNP as "'neutral' territory, where man and beast could peacefully live with each other." He went on to write that grizzlies and other wildlife species "discovered" that the park provided them with a "safe refuge," such that "the [bears] soon learned the boundaries of this unfenced [p]ark, [and] they show[ed] a different nature within its sacred limits. They no longer shun[ned] the face of man, they neither fear[ed] or attack[ed] him, and they [we]re even more tolerant of one another in this land of refuge."[3] For Seton, by producing or enhancing "fake" natural histories like these, his readers were urged to see park boundaries, where wildlife must behave "nicely" to people, as an appropriate fee imposed on wild animals living in a "safe refuge." This idea of YNP as a "peaceful realm where nature-loving humans and reasoning bears observe[d] a truce to never harm one another"[4] created a very specific public perception that was also promoted by William Hornaday, an early conservationist, who believed that protected bears living in proximity with people would never "break faith" with them. Even though he wrote enthusiastically about the high "moral character" of YNP's bears, he saw nothing wrong with physically punishing or killing offending bears "to teach them to keep their place."[5] If bears behaved well according to human expectations in the bounded park, they apparently knew their place and so were "good" bears, and if they ignored their place and behaved like wild animals, they were punished by being roped, tied, and whipped or shot because they were "bad" bears.

Defining what an animal's place was came through the lens of metaphors and stories. For people, grizzly bears eating human food waste was appropriate while killing a horse or acting aggressively around people was not. Bad bears would be called "man-eaters," "cattle-killing bears," or "renegade bears," such that these "bad" bears were the dangerous ones while "ordinary" grizzlies were not. This dichotomous view led park visitors to act foolishly around the park's top predators, and it encouraged management practices that not only maintained a fictional understanding of grizzly bear-human relationships, but by feeding the bears trash, also harmed the bears' health and well-being.

Even though by 1970, when the last YNP trash dump was closed, the earlier idea of a renegade bear or what people today call a "conflict bear" has persisted. In the introduction to Ernest Thompson Seton's book *Wahb: The Biography of a Grizzly* Wyoming historian Jeremy M. Johnston highlighted the role of history and "lore" in Seton's "literary creation" of a renegade grizzly bear called Wahb. Johnston's words remind us of how history and those stories persist, such that sports magazines continue to publish survival issues with magazine covers that capture the battle between a ferocious and vicious-looking grizzly bear, with claws and dangerous teeth exposed, ready to maul a hunter. Popular literature, as well as television shows and movies, have relied on gruesome human deaths by a grizzly bear to highlight how dangerous these bears can be to "innocent" human victims. To make this point, Johnston used the 1966 film *Night of the Grizzly,* a film that follows Clint Walker as a new rancher who must deal with an "outlaw grizzly" named Satan that is killing his livestock. Other hunters try to kill the bear, but Walker eventually succeeds, bringing "peace and prosperity" back to the town. The 1976 film *Grizzly* recounts the violent killings by a "renegade grizzly,"[6] while the 2015 film *The Revenant* portrays a grizzly bear that, for over six long minutes, rips and slashes Leonardo DiCaprio's throat in a gruesome attack. The metaphors of a grizzly bear as a renegade, man-eater, cattle-killer, or conflict bear forge a way of seeing these bears through a lens rooted in human behavior and values rather than the behaviors of a wild grizzly bear.[7] The persistence of these metaphors makes the possibility of human-grizzly bear conflicts the rule for many ranchers and hunters rather than the exception. Evidence to contradict this rule includes the following: the average number of grizzly bear attacks in the entire Greater Yellowstone Ecosystem (GYE) was only 5.45 in 2022, and for YNP that can see four million visitors per year, there were seven people killed by grizzly bears between the years 1872 and 2018. According to Kerry Gunther in her soon-to-be-published article in the scientific journal *Human Wildlife Interactions*, drowning, falling off cliffs, and suicide kill significantly more people than bears.[8] From this, political and agency management decisions and public perceptions distorted by stories, myths, and social cliches to keep grizzlies in their "place" have become the reality for many grizzlies living in the GYE.

BISON, CARIBOU, WOLVES, AND BEAVERS

To understand how stories and metaphors have built human-nature boundaries, the cultural and natural histories of wildlife species must be told. To understand what imagined wildlife means to species like the bison, woodland caribou, gray wolf, grizzly bear, and North American beaver, this book will show how the discourse forged on anthropogenic landscapes has created and maintained very specific human-nature relationships. The gray wolf had no place in the rapidly developing United States, while for the American bison, North American beaver, and woodland caribou, the expectation was there would always be more of each of these species over the next ridge or in the next wetland. The placeless nature of gray wolves made extermination of the species seem like the only option for those living in wolf country, while the impossible idea that beavers, caribou, or bison could disappear made killing them to make money or room for cattle a logical solution in a world that prioritized human spaces over those of the wild.

Conversations with park rangers, ranchers, hunters, wildlife biologists, tourists, and others I met on these human-defined landscapes gave me tremendous insight into the power metaphors and stories have played in creating ways of separating human communities from the natural world. For some individuals, the highly charged aspect of this topic required them to speak to me anonymously, while others were more than willing to either share their distress over the slow pace of change away from the worldviews these metaphors and stories forged or voice anger that they still resonated at all. From my many conversations, as well as time spent in local communities that included visits to popular parks, cafes, and conservation or species recovery centers, attending wildlife management meetings, hiking bison "killing fields" or a caribou bog, scrambling through beaver wetlands, sharing a paddle in a wilderness area, or reading local news and magazine articles, made choosing the American bison, woodland caribou, North American beaver, grizzly bear, and gray wolf as my case studies for this project easy ones to make.

Although this book is not about Native Americans, First Nations, Métis, or Inuit specifically, I do include some of their stories about and interactions with the natural world to highlight the chasm that developed between their worldviews rooted in complex forms of coexistence with nature and the worldviews of European settlers who saw themselves as both dominant and separate from nature. Where the European settler tended to see interactions with nature and subsequently the stories that evolved from those interactions as linear or as a path to a profitable end, Indigenous inhabitants of the land relied on a more circular relationship where their survival relied on including wildlife and nature in their lived experiences. The stories and metaphors that arose from that linear relationship rooted in domination and separation from nature have made coexistence today a challenging one for many people to accept.

Each of the wildlife case studies I present provides critical insight into how language forms boundaries and from that assigns a species its "place" in our imaginations as well as our lived experiences. The American bison that once roamed over thirty million strong across the Great Plains is now defined as livestock and in most cases lives behind a fence. The woodland caribou that supported First Nations and Native American people with food and fur as well as served as an indicator species for boreal forest health is currently absent from the lower forty-eight states and is struggling to survive in southern Canada. Their status as the gray ghost made them invisible to too many of us so their rapid decline was nearly missed. The gray wolf along with the grizzly bear highlights our animus toward predators that can outcompete us for resources as well as threaten our status as the dominant terrestrial species. From this, their wildness became such a problem that calling wolves vermin or grizzly bears ferocious helped inspire a contempt or fear that encouraged people to keep those species in their "proper" place. For the North American beaver, money from the sale of beaver fur and castoreum led to their near extirpation; it wasn't until two Canadians realized that killing off such an important resource wasn't the wisest social, economic, or ecological move. Through their efforts, the beaver later recovered enough to ironically earn the label of pest or river rat. In Britain, even though beavers had been gone for hundreds of years, Queen Elizabeth II was still receiving two beavers from the Hudson Bay Company (HBC) each time she set foot in Canada. In 1990, the last two descendants of the HBC's obligation were displayed by the Beaver Scouts at the Gateshead Garden Festival in England, where they were given the disturbing names of Adolf and Eva.[9] Metaphors, names, and stories like these helped forge a shared worldview that decided where animals could live or whether they could live at all.

Chapters 1 and 2 introduce how the discursive processes of metaphors and stories have not only created ways of seeing wildlife but have also maintained a human-centric world that exists separate from a biocentric one. Chapter 3 explores the American West, where the American bison (*Bison bison*) was nearly driven to extinction through wholesale hunting. Chapter 4 delves into why one subspecies of caribou, the woodland caribou (*Rangifer tarandus caribou*), has been extirpated from the lower forty-eight United States and is threatened in many parts of southern Canada. Its decline forces us to ask: What happened to the woodland caribou? Chapter 5 considers how the natural and cultural histories of the gray wolf (*Canis lupus*) help us to understand how wolves shifted from being human companions to vermin and how the "ferociousness" of grizzlies nearly got them extirpated in the West, while the final case study in chapter 6 looks at how forty years ago it would have taken four sticks of dynamite and a 12-gauge to get rid of a beaver (*Castor canadensis*), yet today in many places our relationship with the beaver is far less adversarial and significantly less violent (see figure I.1). What changed to shift the language as well as the boundaries that defined the beaver's place for most of the seventeenth to twentieth and the early years of the twenty-first centuries from one of commodity to pest to some form of coexistence?

The book concludes by considering the question of social carrying capacity. Social or cultural carrying capacity is the number of a species people can tolerate in their shared spaces. As wildlife recovers in some areas, the number of recovering animals can outpace a community's ability to adapt. In other places, translocations or reintroductions might bring a species long absent back into people's consciousness as well as their daily lives. The return could require behavioral changes from those individuals who share that space, where park grounds or beaches might be closed for bears during their pre-denning period called hyperphagia or for turtles or birds during their nesting seasons, or even because it is mating season and bull elk are in rut. Other changes might include land-use restrictions because bald eagles are nesting in a backyard tree or an expectation that your domestic sheep and hungry wolves can and must coexist. Behavioral changes to accommodate species long absent in our lives can be challenging, particularly for those who don't even want them there. Yet by changing how we talk and think about wildlife, it can become easier to acknowledge that we are actually part of one interconnected biological community made up of humans, animals, and plants.

Figure I.1 A Beaver in Grand Teton National Park. *Source*: Barbara K. Jones.

NOTES

1. Berry 2002: 20
2. Mohr 9.
3. Johnston 20.
4. Johnston 35.
5. Hornaday 1904: 37.
6. Johnston 45.
7. Johnston 41–46.
8. Arnold 11A.
9. Gow 65.

Chapter 1

Narratives as Discourse

Is It Nature or Nurture?

In 1933, when Aldo Leopold wrote that "[o]ne of the anomalies of modern ecology is the creation of two groups, each of which seems barely aware of the existence of the other," he was referencing how siloed those who studied human communities were from those who studied plant and animal communities. This siloing maintained the idea that humans existed outside of nature. For us to become part of nature, Leopold believed that "[t]he inevitable fusion of these two lines of thought w[ould], perhaps, constitute the outstanding advance of the present century,"[1] yet, sadly, nearly one-hundred years later that fusion has yet to occur.

To explore why Leopold's challenge failed, it is critical to understand why human communities and plant and animal communities have not been studied as parts of an interconnected whole. The commonly accepted explanation is that humans are not part of nature. Environmental philosopher Christopher J. Preston saw this partition between the human world and the natural world as a response to not only nineteenth-century Europeans suffering from hunger and disease but also as a byproduct of the scientific and industrial revolutions. The era of Romanticism took it a step further by defining wild spaces as separate from human spaces. According to Preston, a dichotomous relationship with the natural world developed, where "the place for wild beasts was absolutely not the place for people, [just as], the places people lived were off limits to wildlife."[2]

This discourse of separation from nature is a powerful one that fits in with Mary Rosner and T. R. Johnson's work on rhetoric. Rosner and Johnson, English professors at the University of Louisville and Tulane University, respectively, in their article "Telling Stories: Metaphors of the Human Genome Project," connected the tales of Victor Frankenstein (seeking the secret of life), John Franklin (searching for the Northwest Passage), and

Maurice Sendak (writing of Max who becomes King of the *Wild Things*) as stories embedded in our culture that "seduced" men into believing they had "mastered nature." In analyzing the rhetoric of those working on the Human Genome Project, they determined the project scientists relied on three compatible metaphors to describe their work. The metaphors "nature as a book or library," "nature as a machine," and "nature as the unexplored country" placed the scientists in the roles of collector and interpreter, mechanic, and explorer that in each dominated nature and exploited it as the "other."[3] By othering not only wildlife but also their habitats, they and their ecosystems become distinct and less familiar to us. With each degree of distinction, our understanding of a shared world made up of human communities alongside plant and animal communities becomes not only harder to process but also overwhelms already established taxonomic categories. These distinctions also make it more difficult to align an ecosystem's biological carrying capacity to our social carrying capacity for wild plants and animals.

For Jozef Keulartz, associate professor of applied philosophy at Wageningen University and Research Centre in the Netherlands, metaphors have a threefold function. As a heuristic device that can act as a mental shortcut, they allow us to conceptualize unfamiliar ideas and knowledge (thinking function), they facilitate communication across groups (talking function), and they give us paths for decision-making (acting function). As a tool to help people conceptualize new ideas in specific ways, metaphor pairs like wilderness as chaos or wild boars as plunderers can encourage one way of thinking, while wilderness as the sublime or wild boars as woodland regenerators can encourage another. A well-designed set of word pairs can reaffirm, challenge, or alter someone's preexisting mental categories. In 1981, when William R. Jordon III introduced the term "ecological restoration," where degraded ecosystems were returned to their original state, the term had been interpreted in terms of art restoration. For those who saw art restoration as an example of feigning the original, the metaphor pair implied ecological restoration was "Faking Nature" or "[A] Big Lie" because restorations were not authentic so restoring nature was just a way to "disvalue and dominat[e] nature,"[4] while others who saw art restoration less as artifice but more as a process, responded to the metaphor through the lens of artistic recreation where restoration led to enhanced ecosystem functionality. To use metaphors like these to reaffirm or overwhelm preconditions, embedding them in stories can fast-track an already abbreviated process.

This preconditioning is a critical aspect of how metaphors and stories work. Since metaphors act as information shortcuts, where people draw on what they already know to better understand the world, they allow for what American psychologist Keith J. Holyoak and Canadian philosopher Paul Thagard call an "intelligent transfer of knowledge."[5] This transfer of knowledge through metaphor allows individuals to use what is familiar to act as a base equivalent or analog that can be overlaid on new information. In anthropology, this knowledge transfer is processed through preexisting layers in a person's mental map of reality or worldview. By relying on familiar taxonomic categories, the idea of seeking out the familiar in the foreign can reduce someone's ethnocentric tendency to other the unfamiliar. However, since metaphors act as mental shortcuts, they can potentially bypass a "weaker" familiar and skip directly to a "stronger," more established familiar. The language of coexistence shows us just how easily the metaphor's ability to shortcut knowledge transfer works. Like the example with art restoration, the more powerful pathway allowed some people to process it through the lens of faking nature, while others to see it as enhancing ecosystem functionality. To be a truly effective metaphor, it must find mental shortcuts to a shared familiar across more stakeholder groups. If, through cognitive bias, the shared familiar places humans outside of nature, metaphors about coexistence might find little fertile ground for knowledge transfer.

For Leopold, since metaphors and stories are forged in human communities, they can easily be designed to place plant and animal communities outside that human space. The persistence of such a separation has allowed a human-centric "familiar" to become, for many, their mental maps' most powerful imprint. For many human communities, an unwillingness to find or seek out patterns or commonalities that would shift existing shared worldviews toward a more inclusive biocentric "familiar" has helped maintain a dominance-over-nature relationship with the natural world. This dominance metaphor lets human actions act as a hegemonic force that can deny its animal or plant victims the agency to satisfy their needs and wants. For wild species, living in plant and animal communities distinct from human communities can be an isolating and limiting space to occupy.

NATURAL DISCOURSE

Work done by John Dryzek on rational ecology shows us that the notion of discourse can be useful in creating pathways for understanding that offer some agency to competing viewpoints. Throughout life, each of us develops a set of assumptions, judgments, and contentions that provide the baseline for dialogue and can determine whether we are able to "see" or "hear" what someone else is saying.[6] The more vested we are in the themes associated with a specific discourse, the more difficult it is to appreciate another person's point of view or see any coherence or logic in their stories. Conflicting assumptions, judgments, or worldviews create discourse boundaries that are hard to overcome and can help explain why people see nature in such different ways.

By admitting that divisive discourses do exist and can create controversial disconnections, we can begin to think about how they are formed and how they persist. To Dryzek, there are four elements that contribute to our specific worldviews and "enable stories to be told"[7] that reinforce our existing assumptions and contentions. The most obvious is that human behavior isn't the same everywhere, in that people operate from within different norms, institutions, and symbolic meanings, so they are not going to conceptualize things in a similar fashion. His next element is recognizing that people see natural connections and relationships differently. Someone who sees a wolf as vermin is going to see natural relationships differently than a person who sees the wolf as the return of the wild. Element three considers the motivation of the different players. Those who feel their lives are negatively impacted by the presence of prairie dogs or bison will see those species in a different light than a visitor hoping to see bison in Yellowstone National Park (YNP) or prairie dogs in Wichita Mountains National Wildlife Refuge in Oklahoma. Those visitors might have traveled great distances to witness wild nature and experience the joy of watching a bison wallow or a prairie dog whistle, while the rancher might have had to deal with the economic losses associated with his cattle having to compete with bison for grass or the worry over a cow potentially breaking its leg after stepping into a prairie dog hole. His final element addresses the importance of metaphors as rhetorical devices that convince others to see a situation in a particular way. An individual who sees wilderness as a metaphor for the uncontrolled and unregulated might have a hard time appreciating wilderness as a "pristine ideal" or value it because it is "untrammeled by man."[8]

Discourses can in part be understood and analyzed by considering the different metaphors people use to understand the world. This occurs because the way individuals think and act is determined by his or her own conceptual system, which in turn is defined by metaphor. According to linguists George Lakoff and Mark Johnson, "the essence of metaphor is understanding and experiencing one kind of thing in terms of another."[9] For Karis Moser in her 2000 article "Metaphor Analysis in Psychology—Method, Theory, and Fields of Application," metaphors, because they both operationalize tacit knowledge and mirror social and cultural processes of understanding, give us a way of seeing a person's thoughts and actions,[10] while to archaeologist Christopher Tilley, such an understanding occurs because metaphors are culturally specific. They allow an individual to see something in a certain way because that something is filtered through a "culturally mediated bodily experience."[11] Unlike other imaginative language, metaphors rely on the individual participant to do the work of establishing the connections and constructing a particular logic that is involved in the transfer of meaning.[12] For nature enthusiasts, the experiences that allow for this transfer of meaning might come from exposure to nature or learning to see nature as valuable in both intrinsic and extrinsic ways. For the environmental community, these experiences vary depending on the agency or organization's objectives regarding environmental protection, while for the landowner or highway planner, by costing money and time, living with nature can be a nuisance or a burden they are not willing to bear. To Mark Vieira, the Carnivore and Furbearer Program Manager for Colorado Parks and Wildlife, circumstances can also allow a negative or positive response to the presence of, for example, a beaver as temporary in that most people "can hold both ideas [about the species] at the same time." To him, different circumstances may result in different points of view so nuisance and admiration can be flip sides of the same coin.

For a farmer or rancher, the metaphor "pastoral ideal" or "kindly use" allows them to conceptualize their everyday reality in terms of agriculture and country life as a model of social, economic, and environmental excellence, while the metaphor "pristine ideal" or "pristine wilderness" encourages the environmentalist to conceptualize natural areas as pure and uncorrupted versions of a people-less wild. Agriculture by its very nature cannot possibly fulfill the environmentalist's vision of the pristine ideal, just as the pristine ideal makes no sense to the farmer who relies on his vision of the pastoral ideal and the kindly use of his land to make sense of his world. Here, where environmentalists look to preserve nature for its own sake, farmers or ranchers, by relying on what they see as the wise management and efficient use of resources, look to work with nature.[13]

This dichotomous way of seeing the land has had a long history with Aldo Leopold in 1925 arguing that wilderness as land free from humans was a relative condition and couldn't be seen as rigid and unchanging, but instead, "must be a flexible thing, accommodating itself to other forms and blending with them."[14] This fits in with writer and conservation biologist Curt Meine's defense of the "relative wild," where there is a "continuum of land-use intensity" that argues against both the "myth of the humanized" and the "myth of the pristine."[15] The failure to understand our relationship with nature through this more relative and flexible lens can forge an imagined geography defined by human-imposed boundaries that continues to put us outside of nature.[16]

FOUNDING MYTHS AND STORIES

The American Frontier Thesis, initially espoused by Frederick Jackson Turner in the 1890s, became a way of seeing America. That myth relied on the idea that the frontier of the Midwest and Far West would forge a shared

American consciousness separate from its European roots. That consciousness would allow Americans to become their true selves, one that relied on self-sufficiency, pragmatism, egalitarianism, and civic-mindedness. Turner believed the frontier, with its "uniquely American character," would take America's immigrant mix and forge it into a nation of united people. To Turner and subsequent historians, the presence of this frontier aligned with America's diversity and its reliance on scientific observations rather than older ideas about being a chosen people to help explain the feeling Americans had about their nation's exceptionalism. Less than ten years later, Turner realized he was wrong. The shared American character that he thought would flow from people moving West toward the frontier did not occur; instead, people remained closely anchored to earlier "settler streams" of the Scots-Irish in Appalachia, the British Protestants of New England, or the broad range of people who made up the American South. So even when they moved from Appalachia, New England, or Mississippi to destinations west, they took their worldviews with them. These different worldviews, instead of blending together to forge a new American worldview, remained somewhat separate and distinct across the nation. The regional differences or sectional ties that existed in America and resulted in part from the diversity of settler streams were not overcome by the presence of shared ideals and government institutions and laws.

Research done by Colin Woodard in his 2011 book called *American Nations: A History of the Eleven Rival Regional Cultures of North America* offers us insight into why Turner's expectations regarding settler streams in a world of shared ideals didn't happen. To Woodard, the red and blue partisan divide that exists across America is expressed today in how regions deal with issues like healthcare, gun control, or life expectancy and can be understood through the settlement patterns that formed what he identifies as eleven North American cultural regions. Yankeedom, which includes eastern Maine to New York with some overlap from New Netherland and the Midlands, was founded by Puritans who worked to make the earthly realm perfect through social engineering, that chose the common good over individual liberty when the two were in conflict, and accepted outsiders through assimilation, while New Netherland founded by the Dutch to be a global commercial trading culture, was multicultural, committed to tolerance, and the freedom to think and make one's own value choices. Greater Appalachia (Pennsylvania, Virginia, North Carolina, western South Carolina, and eastern Tennessee) was developed by settlers from war-torn places in Northern Ireland, Northern England, and the Scottish Lowlands. They were committed to personal freedoms and had a well-developed suspicion of outside authority. The Left Coast (western parts of California, Oregon, and Washington, as well as eastern Alaska) was a blend of New England merchants and missionaries who left Yankeedom and midwestern Appalachian farmers and fur traders. Those from Yankeedom brought their ideas of utopianism, and those from Appalachia brought their need for personal sovereignty and a willingness to explore. That push to create a western Yankeedom continues to clash with the Far West (the states of Montana, Idaho, Wyoming, Nevada, and Utah, and parts of Alaska, Washington, Oregon, California, Arizona, New Mexico, Colorado, North and South Dakota, Nebraska, and Kansas), or what Woodard calls the "second-generation" nation.

Land west of the one-hundredth meridian, in the Far West, is higher and drier, making traditional eastern-style agriculture that relies on more predictable rainfall and better soil nearly impossible. That means instead of growing foodstuffs, the Far West has had to rely on industries that included railroads, mining, ore smelters, and equipment production, as well as the loss of millions of acres of western land that was being set aside for national parks and monuments. The western mountain land that most people believed would be open to private individuals, as had been the norm, was now public land controlled by the federal government or the railroad. Like Appalachia in the early twentieth-century coal-producing years, the Far West saw itself as an internal colony within a larger country. To Woodard, a sense of victimhood developed as decisions about how the region would be settled, who controlled the land, and how the region's extracted wealth didn't stay in the West but went to the federal government and corporations headquartered on the east and west coasts. This feeling of exploitation by powerful outsiders resulted in a resentment that today is directed toward the Left Coast and the federal government. The resentment of the Far West, along with the optimism and appreciation for the common good expressed by those of Yankeedom, can help explain why Turner's settler streams didn't turn into one cultural region with shared ideals. Today, Montanans and Wyomingites see the unpopular changes to their states from the Left Coast as the "Californication of Montana," and the overcrowding in Bozeman, Montana has turned it into "Bozangelas." When a trash ordinance was proposed in a small Wyoming town, the angry residents felt it legitimate to yell at the town's leaders, "This Is Not California!" The language of Californication and Bozangelas, as a reference to the large population and urban sprawl of California's Los Angeles, reinforces a story about the "Left Coast" California at odds with the "Far West" Montana or Wyoming that hasn't encouraged the formation of shared ideals. To Turner, those shared ideals would have allowed for compromise and an environmental ethic that saw land stewardship rather than starting somewhere fresh as the only option. For this book's purposes, the stickiness of settler streams in Turner's Frontier Thesis can help explain the persistence of old stories and myths against the emergence of new ones as explanations for how people see human-wildlife relationships and forge ideas about wildlife's "place" in those relationships.[17]

REWRITING THE STORY

A lack of regard and at times intense dislike for specific animals can occur when the borders of our human-defined imagined geography are crossed. The border that separates highly controlled human spaces of "clean [and] orderly civilization" from "wild, uncontrollable nature" is intended to be inviolate; so when boundary crossings do occur, we are forced to recognize that our controlled human spaces are not as free of nature and its unpredictability as we imagined.[18] Wolves that enter the domain of the rancher or bald eagles that feed on a farmer's chickens have in each instance crossed that inside-outside border we have created in our minds. Wild, then, is the ultimate foreign experience for us. According to Irus Braverman in her work *Wild Life*, wild is not confined to the animals themselves[19] but includes the setting of their existence and the "context in which they evolved."[20] What culture and place do for human identity, the evolutionary pressures found in the wild do for wildlife. Wild is based on a set of behaviors we can't always predict; so when we enter those wild spaces, we tend to be unprepared and at times ignorant of the motivations that drive them. Not that long ago, a female visitor at Custer State Park in South Dakota got too close to a wild bison calf and suffered the consequences when the mother bison charged and caught the woman's belt in its horns. She was violently swung and saved only when her pants came off and she fell to the ground unconscious. This woman failed to recognize that in this instance she was the boundary crosser, and through her actions had entered the relative wild. To minimize such negative human-wildlife interactions, YNP has posted signs about keeping safe distances and behaving appropriately around wild animals, with one effort at humor that reads "don't pet the fluffy cows" to get people to understand the park is not a petting zoo. When visitors enter La Chua Trail in Paynes Prairie State Park in Florida, they encounter a bold yellow sign that reads: "Entering Wilderness Area. Use Extreme Caution. Remember this is NOT a zoo or theme park." These signs are an attempt to educate and remind visitors they are leaving the predictable space of their imaginations and entering a landscape that is the setting of wildlife's natural existence; yet the ineffectiveness of signage can in part be explained because people behave through past mediated experiences that come from outside any specific event or request. People will get too close to bison, take photos of a grizzly bear gorging on a carcass while they are watching from thirty feet away (the distance a grizzly bear can cover in seconds), or pose next to a sleeping alligator because they want a close-up that records their fearless dominance over a wild animal. To document as well as poke fun at the inappropriate behaviors people engage in at YNP and Grand Teton National Park (GTNP) there is now a site on Instagram called "touronsofyellowstone." The word "touron" comes from blending tourist with moron and is used to identify park visitors who are forced to run from bison, or, according to one commenter, can "almost [get] smacked" by a black bear because they disregard the one-hundred-yards-from-predator message. The less control we have, the more wild and unpredictable nature appears to be. Like Meine's description of relative wild, to the Center of Biological Diversity, wildness is a point on a continuum that tolerates different degrees of "human manipulation."

Coexistence as the metaphor for our relationship with wild nature rests on the assumption that this human-nature relationship is shared. Yet for many, coexistence conflicts with the equally powerful narrative of dominion over nature. Pest removal companies offer landowners or ranchers ways to remove prairie dogs, a species that improves habitat and provides shelter for other species, yet dominion-over-nature marketing ads convey the message that prairie dogs can "destroy a yard's beauty." In Newfoundland, even though sea otters are essential to maintaining balance in coastal marine ecosystems, they are called rats by fishermen because they break into their lobster traps and eat their catch. To one Newfoundlander, fishermen kill the otters because, like non-native rats, they are pests, while squirrels that are also called rats are shot because they tear up empty cabins, chew up wires, and rip upholstery. The last in his list of pests is the moose that had been introduced to the island in 1906, but today their presence has become "too much" because they are road hazards. The dominion-over-nature narrative is a powerful message that resonates across different stakeholder groups, making coexistence less likely when pest metaphors maintain a more familiar worldview.

A powerful way to achieve that pull without the push is to rely on storytelling. Work done by political theorist Hannah Arendt reminds us of this when she writes, "Storytelling reveals meaning without committing the error of defining it,"[21] thus letting "egos sleep."[22] Since stories are not imposed on listeners, they are an ideal tool for engaging decision-makers because a good story lets decision-makers find their own meanings. Deciding when to tell which story to generate the intended meaning is where things get tricky. Robert Chaney, author of *The Grizzly in the Driveway*, noticed something similar during the federal delisting process for two highly polarizing, yet endangered species: the wolf and the grizzly bear. At the delisting hearing he attended in 2010, the wolf generated a crowd of wolf lovers and wolfers who were easily identified by the wolf toys one group was clutching and the orange hunting vests the other group was wearing. Eight years later, while at the grizzly delisting hearing, there were no hunters present, only grizzly enthusiasts. When Chaney asked where the grizzly hunters were, a lawyer responded, "It's the difference between protecting Little Red Riding Hood and shooting a Teddy Bear."[23] Little Red Riding Hood was saved by the hunter who shot the wolf, while the Teddy bear became a childhood icon. For bears, that was a good story told at the right time.

Today's challenge then is to forge stories and metaphors that allow more of us to see logic and coherence in a less contentious relationship with nature. Such a relationship can help shift the linguistic boundaries we rely upon to keep us separate from wild nature into more porous ones that embed us in nature. The effects of climate change might make this pitch an easier story to tell because healthy habitats and intact ecosystems are much more resilient to fire, drought, and flooding. When Aldo Leopold wrote that "[t]he inevitable fusion of these two lines of thought will, perhaps, constitute the outstanding advance of the present century," he was challenging us to understand that our human communities are indistinct from animal and plant communities. Recognizing that nature's well-being and our well-being are interconnected offers us a powerful response to that challenge.

NOTES

1. Leopold 1933: 182.
2. Preston 28.
3. Rosner and Johnson 104–105, 107, 109, and 115.
4. Keulartz 33.
5. Holyoak and Thagard 1995.
6. Dryzek 8.
7. Dryzek 15.
8. The Wilderness Act 2.
9. Lakoff and Johnson 5.
10. Moser 4–6.
11. Tilley 45.
12. Tilley 45.
13. Cochrane 1993; Lakoff and Johnson 1980.
14. Meine 90.
15. Meine 90–91; Jones 20.
16. Said 1979; Braverman 2015.
17. Woodard 22–24; Cronon 1987: 165–170.
18. Soniak 11/14/16.
19. Braverman 44.
20. Braverman 105.
21. Arendt. Universal Storyteller.
22. Arendt. Universal Storyteller.
23. Chaney 193.

Chapter 2

Reconsidering the Boundaries Wildlife Stories and Metaphors Have Built

In a special annual conservation issue produced by *Jackson Hole News & Guide* called *Headwaters*, Susan Clark of Yale University's School of the Environment wrote how the stories that maintain a sharp divide between us and nature are "outdated, unproductive, and not realistic."[1] These stories we have come to rely upon to define and describe wild nature evoke a world that no longer exists. The linguistic tricks that made dominion over nature an appropriate way of seeing our relationship with nature have led to the extinction and extirpation of species, the pollution of water and air, the destruction of habitats, and the redefining of every species' evolutionary relationship to their environment to one managed by us. The metaphors and stories that forged this association have relied on branding and assigning nature value that are long past their sell-by date.

The metaphor "hunting is conservation" gives humans the role of top predator in habitats and landscapes where native wild predators are no longer welcome. The reintroduced Tule elk in the Point Reyes National Seashore in Northern California rely on humans to manage their numbers whenever they exceed the carrying capacity of their small, tightly constrained landscape. The elk's wild predator, the California grizzly, has been absent from California since 1924 when the last bear was shot and killed, so because grizzlies are no longer welcome to share that landscape with ranchers, homeowners, and seashore visitors, their absence leaves us to serve as the replacement in wild elk population control. Hunting as conservation gives hunters and trappers tremendous credit for effective wildlife management, yet evidence from the near extinction of the American bison, the extirpation of wolves, bears, and elk from regions throughout the United States, and the extinction of species like the passenger pigeon suggests otherwise. Many hunters kill to get a trophy or with little regard for species' viability and, without a myriad of hunting rules and regulations, cannot be depended on to value wildlife through the lens of conservation. While hunting organizations like the Rocky Mountain Elk Foundation, Muley Fanatic Foundation (for mule deer), Ducks Unlimited, and even the Boone and Crockett Club restore habitats, buy tracks of land, and provide financial backing for many conservation projects, their ultimate endgame is to increase habitat for game animals. Since hunting is deeply rooted in many of the outdated and unproductive narratives that have sharply carved us out of nature, the "hunting is conservation" maxim is not necessarily a reliable one for defining our relationship with the natural world.

Although the California state flag depicts a California grizzly, they are not welcome to live within California's managed spaces. That conundrum presents itself in places throughout the world—over eight billion people live on this Earth, and, in many instances, they find cohabitating with what they see as pests, vermin, or dangerous beasts highly problematic. This not-in-my-backyard (NIMBY) issue reminds us that people don't all see coexistence in the same way or share the same coexistence stories and metaphors. A coyote in New York City's Central Park might be a novelty in the short-term but what happens when it runs too close to a child in a stroller or chases a little dog in a dog park? A fox taking up residence in a small suburban wild nature area might be great to talk about at the school bus stop, but what happens when the fox roams through the school playground on its way home? In both instances, the animal likely will be removed because it has crossed the imaginary "backyard" boundary that carves humans out of wild nature.

Our relationship with wildlife has been defined by stories that vary in their inclusiveness. Even early naturalists appreciated wildlife through very self-serving lenses that operated within different versions of man's dominion. Efforts to conserve nature by John Muir, George Bird Grinnell, John Audubon, and Theodore Roosevelt didn't push people to toss the dominion narrative; instead, they altered what dominion meant. Such a fluid definition rationalized Audubon's love of birds with his opinion that wolves ought to be eradicated for the threat they posed to valuable livestock or for Muir to not only see Yosemite as the wilderness ideal, but to run a sawmill at the base of Yosemite Falls for developing tourist infrastructure. Muir also took joy in the "militarization" of parks using armed guards to enforce rules and manage the ecosystem to exclude sheep and prevent fire.[2] When Muir complimented the valleys in Yosemite by calling them a "great landscape garden," he was using that language to explain how Yosemite was unlike the formal tended garden yet still could be understood through the lens of a garden metaphor. Such dominion stories that rely on garden or livestock-over-wolves metaphors persist due to their wide umbrella appeal to a larger constituency.

For coexistence to become the relationship we have with nature and wildlife, the long-lasting and persistent "dominion over nature," "hunting is conservation," and "ranching as a way of life" soundbites need to be recognized as outdated and narrowly constructed truths. Today, coexistence as a way of seeing human-nature relationships is fragmented, and our relationship with many wildlife species relies on tautologies that reinforce old stories, myths, and human-nature relationships. As we rethink these stories and myths about wildlife and nature so that maintaining our mutual well-being can occur in our own backyards, new stories and modes of discourse must seek to reimagine and revalue nature so that those tenacious assumptions rooted in the past can lose their influence.

DOMINION OVER NATURE

For the white Europeans who settled in North America, dearly held ideas about private property, market economies, and the productive use of nature traveled with them and formed their relationship with America's natural bounty. This relationship was defined through the lens of domination and played out in how these newcomers interacted with both the existing inhabitants and the country's abundance of natural resources. Where America's Indigenous people typically saw the land and its resources through an ecological lens that recognized their interrelationship with it, the new settlers, with their own ideas on property rights and market economies, saw nature's bounty ripe for making money. A lack of rules or a market trigger to manage this bounty in a sustainable way let farming, deforestation, and mining, as well as wholesale hunting and trapping, become the norm. These failures led to the loss of many species and their habitats.

Garrett Hardin called this outcome a tragedy of the commons, where resources not assigned value or ownership are by default held in common for all of us and from that open access risk being lost to us all. Common access typically provides no incentive to treat the resource as an asset for the future, but rather something that can be depleted by the last user. Without cultural institutions or government rules to impose ownership or assign value, a tragedy of the commons can be the result. Who owned the bison on the plains, the beaver in the wetlands, the passenger pigeon in the sky, or the wolf in the American West? To the fur traders, settlers, ranchers, and frontiersmen, no one did and so bison numbers went from more than thirty million to a few hundred, beaver from 400 million to 100,000, passenger pigeons went from three billion to none, and wolves were extirpated from the lower forty-eight states except for a few that remained in Minnesota and Isle Royale in Michigan. Dominion over nature led to a tragedy that we are still dealing with today.

Following the last ice age, wolves and the earliest humans in North America adapted to this land together in a somewhat mutually beneficial coexistence.

That relationship changed when humans settled down with domesticated livestock. Once we moved to a food-production mode of living, humans and wolves began competing for the same livestock resource, an arrangement that didn't sit well with these early pastoralists. The wolf was just one of the many species that lost its place in the interconnected web of humans and nature. So, while wolves competed with both livestock and game, the bald eagle was accused of eating the settlers' fish and disturbing their livestock. Bad press from Benjamin Franklin and his belief that the bald eagle was a "bird of bad moral character; he doesn't get his living honestly . . . [and] besides, he is a rank coward"[3] contributed to an anti-eagle attitude. This was only reinforced by fishermen's complaints about bald eagles gorging on their salmon.

A bird that in pre-colonial times numbered 50,000 pairs in the lower forty-eight states dropped to only 400 pairs by the 1960s. In Alaska, fishermen would ransack and destroy every eagle nest they found to the point that slaughtering eagles became "more or less an established custom."[4] The National Rifle Association (NRA) further legitimized the practice when they called shooting bald eagles the "purest of all rifle sports."[5] Despite Franklin's disdain, in his day there were tens of thousands of bald eagles in America, making this bird a natural symbolic choice for America. To his peer Thomas Jefferson, the bald eagle was "free-spirited, high soaring, and courageous,"[6] behaviors that matched those of the new nation. Yet, this bird that represented what it meant to be American and was "the embodiment of freedom and power" was being killed at an alarming rate. Our human tendency to "kill what couldn't be dominated,"[7] as well as, in later decades, the ease by which chemicals could dispatch these unwanted "pests" made the world a difficult place for eagles to survive.

As with other species, bald eagles had a bad press agent. They were accused of killing everything from salmon to young reindeer, leading one biologist to say in 1929 that "forty thousand [bald eagles] could not appreciably affect the supply of salmon in forty thousand years. But man, with his wasteful methods, intent only on his wasteful gains, must find a culprit and the eagle was a convenient suspect."[8] Like the wolf, the tendency of the bald eagle to eat food shared by humans, whether it was an occasional salmon snack or a quick raid of a farmer's lot, encouraged people to blame them for more than their share of crimes. Some people believed eagles could even carry away human babies. There was so much credence given to this tale that in a 1946 brochure, the Minnesota Department of Natural Resources reassured the public that baby stealing by eagles was highly unlikely. Yet, in January 1961, when *National Geographic* sent a photographer out to document a bird that had become all too rare, the feeling was that the emblem of America was facing extinction.[9]

Passenger pigeons are an example of a species that faced human domination and lost. A. W. Schorger, in his 1955 classic study on the extinction of the passenger pigeon, estimated there were three to possibly five billion passenger pigeons in North America at the time of European arrival, while Joel Greenberg, in his 2014 work on the passenger pigeon's extinction, wrote that the population of passenger pigeons might have exceeded that of every other bird on earth. That is a lot of passenger pigeons to lose through the efforts of just one predator. The numbers killed in wholesale hunting were unimaginable, with "one citizen report[ing] to have killed 475 with a single stick,"[10] while other pigeon hunters shot hens on the nest because they claimed the orphaned squabs (young pigeons) would be cared for by other hens. This claim, like the wolf and livestock stories, minimized in the hunters' minds the impact of their indiscriminate killing. One pigeon hunter, William Leffingwell, wrote in 1895, "There is a particular charm about pigeon shooting; it captivates everyone who participates in it. To-day we kill ten straight and rejoice because of it; to-morrow we kill seven out of ten. Now we are dissatisfied and try again."[11] A population of over three billion birds blinked out because of us. Their absence has changed ecosystems in ways we still can't imagine and reflects poorly on our ability to regulate our excesses. Without enforceable rules and values that are shared, the tendency to try to dominate nature for short-term gains seems to be a human default (see figure 2.1).

Our dominion over nature has led to the loss of species as diverse as the Steller's Sea cow and the Carolina parakeet. These extinctions can be blamed on habitat destruction, market hunting, poaching, poor species management, pollution that has contributed to climate change, and the introduction of

Figure 2.1 In an Exhibit at the New York State Museum Called "Gone Forever," the Passenger Pigeon, Carolina Parakeet, Heath Hen, and Eskimo Curlew Are on Display. *Source*: Barbara K. Jones.

non-native species. The totality of this assault on nature has made the sixth extinction our potential legacy to the Earth's natural history.

HUNTING AS CONSERVATION

The North American Model of Wildlife Conservation, introduced by early conservationists Aldo Leopold and Theodore Roosevelt, recognized two key principles: our fish and wildlife belong to all North Americans, and they are to be managed so their populations will be sustained forever. From those principles, a set of foundational guidelines that were called the Seven Sisters of Conservation emerged. Relevant to our efforts at coexistence are the model's principles that recognize wildlife as a public trust resource, that there should be no markets for wildlife, that wildlife can only be killed through fair chase, and its killing must be for a legitimate purpose. Those principles rested on the claim that hunters were the best caretakers of public wildlife. Yet, in the 2012 technical review of the conservation model by The Wildlife Society and the Boone and Crockett Club, the reviewers asked how the widely spread practice of shooting prairie dogs and even the hunting of crows satisfy the question of legitimate purpose? Do those practices fall within the legitimate use principle when those species are not killed for food or protective benefits? These questions were followed by another that asked how tenacious predator removal or control programs fit within the principles of the North American Model?[12] In other words, is baiting a wolf so it will leave a no-hunting zone fair chase, and is killing a wolf just because it is a wolf legitimate use? These questions align with the words of Aldo Leopold on display at the Boone and Crockett's Headquarters in Missoula, Montana: "We abuse land because we regard it as a commodity belonging to us. When we see land as a community in which we belong, we may begin to use it with love and respect."

Like the contradictions with fair chase and legitimate use principles is the language of trophy management areas or trophy hunting, where big buck contests, coyote, wolf, and bear hunts, and shed hunts (where people collect antler sheds to sell) imply the existence of a market for wildlife.[13] Hunting for trophies is problematic because it conflicts with better wildlife management strategies. According to the Maine Department of Inland Fisheries and Wildlife (DIFW) Wildlife Division Director Nathan Webb, for the fall 2021 season, deer hunters were asked to kill more does because hunters weren't harvesting enough antlerless deer even though the agency was issuing more permits for that purpose. The urge to get a trophy buck over other options conflicts with the conservation model's principle of killing wildlife for legitimate purposes that includes food, fur for clothing, self-defense, or protection of property as principle number four. If high deer numbers were being managed as a resource, there would be no permits issued for trophy bucks (3″ or larger antlers) for the 2021 hunt.[14] The trophy management area for wolves relies on the same kind of language and motivation. In Wyoming, this zone is outside Yellowstone National Park (YNP) so wolves may be killed as trophies based on a set quota, while outside the trophy management area a person can kill a wolf for almost any reason. In Montana and Idaho, the practice of snaring wolves in traps, killing them in their dens, or baiting them out of the parks highlights the failure by these hunters to honor the model's fair chase clause.

When the Greater Yellowstone grizzly was being considered for delisting in 2007, the federal government used the language of trophy hunting to suggest a causal link between killing grizzlies and an increase in public tolerance for grizzly bear recovery. In 2018, in the hope that the grizzly bear would be delisted, Montana, Idaho, and Wyoming created a post-delisting grizzly bear management strategy that included hunting to manage grizzly bears as a game animal. Out of fear of court challenges to the delisting rule, Montana did not set a 2018 grizzly hunting season, but Wyoming and Idaho did. Wyoming put twenty-two permits in a lottery system that offered hunting access to ten hunters in prime grizzly habitat near YNP and an additional twelve permits for hunting in less ideal areas. Idaho issued one permit through a lottery and that hunter would be accompanied by a state game warden to make sure no sows were killed. In a curious twist, in the most coveted district, two Wyoming permit winners were photographers who intended to "shoot" the bears with cameras. Hunters were angry over those lost opportunities, but grizzly bear fans were thrilled that two out of the ten permits would not result in a dead bear. After failing to get the Greater Yellowstone grizzly delisted in 2018, in November 2021, Wyoming again pursued the delisting of grizzlies, but this time without Montana and Idaho. With the failure of Wyoming's petition to the United States Fish and Wildlife Service (USFWS) to delist the grizzly bear, in January 2025, the state began considering a new proposal that would attach a free "bear coupon" to all resident elk licenses. The coupon would allow for "heavy grizzly bear hunting" in 2026 and 2027 in areas just outside grizzly bear recovery zones. Even though it would be a federal crime to kill a listed grizzly bear, the proposal would have allowed hunters to kill grizzlies either ten days after delisting or on January 1, 2026, whichever came first.[15] A week later, concerns over violating federal law led to the proposal dying after its first reading on the House floor. Andrea Zaccardi, a senior attorney at the Center for Biological Diversity, said petitions like these are aimed at turning grizzlies into "trophy hunting targets." To her, the past has shown that Wyoming, Idaho, and Montana will do anything for hunting and agricultural interests, indicating they "just can't be trusted to manage grizzly bears."[16]

According to the International Union for Conservation of Nature (IUCN), the "trophy" is a part of the animal that may include the horns, head, or skin or even the shell of a species like the Nautilus, a pelagic marine mollusk, and is usually kept by the hunter or fisher. Shed collecting could be a potentially more sustainable way to collect trophies, but today, where elk and deer antlers were once just "fun trophy finds in the woods,"[17] they are now a source of substantial money. By monetizing the sheds, our relationship with them has changed and moved away from finding antler castaways on the ground as an unexpected discovery to poaching antlers on winter ranges when animals are at their most vulnerable. Other species face similar threats when people harvest them for their trophy value. For the Nautilus, their slow growth and long egg development make their popularity in the shell trade particularly problematic. Their shells are sold as "trinkets," yet there is so little known about them that intense harvest pressure could wipe out a known species or deny us the opportunity to identify an unknown subspecies (see figure 2.2). According to Dr. Peter Douglas Ward, a scientist with Save the Nautilus, there needs to be a cultural moment where people realize we "don't need [a Nautilus shell] to sit in our bathrooms as a trophy."[18] For the grizzly, the trophy could be its claws, teeth, skulls, or pelts, while their gall bladders and paw pads have value outside the normal channels of collection. According to Robert Chaney, though, the biggest prize for the grizzly hunter could be their story, a story that glorifies how the hunter "killed something not only bigger and stronger than [they are], but something known to fight back."[19]

Figure 2.2 Nautilus Shell. Museum of Natural History (Smithsonian). Washington, DC. *Source*: Barbara K. Jones.

Tied to the disconnect between shared users of our wildlife resources is the enduring myth that federal agencies only manage wildlife habitat, while the states manage the actual wildlife. In the United States, wildlife is a public trust resource, which means no special interest groups such as hunting outfitters, ranchers, farmers, or well-connected landowners can influence how wildlife is managed on federal lands for us as its trustees. By allowing states to decide how to manage wildlife within their borders, the federal land management agencies that have an "obligation, and not just the discretion"[20] to manage and conserve fish and wildlife on federal lands have abrogated their responsibilities. This is particularly evident in Park County, Montana, and Teton County, Wyoming, where grizzly bears have easy access to many dump sites. Bears caught taking food from the dumpsters are thought to be habituated and dangerous to people. Traps are set up to capture the bears, which are then either relocated or killed. If the USFWS, state office holders, and wildlife officials were serious about managing the grizzly bear as an endangered species, the dump sites would be bearproofed. To Susan Clark, the inability to fulfill our obligation to wildlife demonstrates a "growing failure of the state fish and game organizations as well as the overall institution of wildlife management and the [North American Wildlife] model itself."[21]

Accepting that hunters and others engaged in wildlife extraction should not bear the entire cost of wildlife management, and that their interests are not necessarily aligned with nongame wildlife protection anyway, under consideration is a way to assign those who are interested in observing wildlife and nature a greater voice. Since no vehicle typically exists for wildlife viewers (from birders to wolf watchers) or ecotourists to assign value to wildlife through their purchases of spotting scopes, binoculars, and associated travel costs, they are left out of the discussion and given no voice in how wildlife is managed. The management of wolves or grizzly bears relies then on those people who might not even want these animals around. In response to that, on April 20, 2021, the owner of Jackson Hole Ecotour Adventures, Taylor Phillips, along with his partners, The Wildlife Fund, Greater Yellowstone Coalition, and the Cougar Fund, created an initiative called WYIdlife Tourism for Tomorrow, where the businesses and people who depend on Wyoming's wildlife "give back" to wildlife. The urge to do this was based on a 2017 economic analysis of the Wyoming Wildlife Recreation Economy that determined hunting, fishing, and wildlife watching generate over a billion dollars in economic activity to the state each year. Wildlife watching alone is responsible for over half of that and employs more than 5,000 people across the state. Since wildlife tourism receives all the benefits of Wyoming's wildlife but currently contributes very little to the conservation of this critical resource, WYIdlife Tourism for Tomorrow has created a funding mechanism for businesses and individuals to contribute to one or more of their carefully selected wildlife projects.[22] This infusion of cash has the potential to realign how species are managed so that the fate of highly polarizing species, like the wolf or grizzly bear, is determined by more of us. The economics of this new funding stream can only help to broaden the role of coexistence in determining how wildlife is managed.

RANCHING OR FARMING AS A WAY OF LIFE WHERE CATTLE IS KING

Traditional ways of living have been successful at maintaining the status quo in human-wildlife relationships. In America's farm and ranching communities, the status quo has been tied to a set of beliefs about farming and attitudes toward farmers, typically premised on an agrarian myth. This myth can be attributed to Thomas Jefferson, who believed that Jeffersonian democracy could best be served by a society made up of farmers. In a 1785 letter to John Jay, Jefferson wrote: "Cultivators of the earth are the most valuable citizens. They are most vigorous, the most independent, the most virtuous, and they are tied to their country and wedded to its liberty and interests by the most lasting bonds."[23]

Because farmers, and subsequently ranchers, were tied to one spot, Jefferson believed that a society made up of farmers was better equipped to see the connection between personal interest and the public good. They had to learn to get along with their neighbors and to take an interest in long-term social stability, so the virtues of honesty, integrity, and charity that promoted community were the virtues that promoted their own interests as well.[24] This agrarian philosophy ultimately saw farming and ranching as the model for unifying our self-interest with the public good (see figure 2.3).

By the nineteenth century, the Jefferson ideal was firmly entrenched in the minds of Americans. The citizen-farmer defined the American national character in that he worked within nature and, through hard physical labor, achieved the virtues necessary for American self-realization. Emerson and Thoreau helped create this romantic image. They were the intellectual leaders behind the movement that redefined human purpose such that society was seen in communion with nature. Yet where Thoreau looked to nature as a place distinct from man's intervention, a place of contemplation, Emerson saw nature as a "place where human beings were busily at work transforming the natural landscape."[25] According to Emerson, farm work brought human needs into harmony with nature's cycles, while farm life was a moral ideal because the virtues of citizenship and harmony with nature were thought to be aligned with farmers' self-interests.[26]

Figure 2.3 Cattle King of Wyoming at Grand Teton National Park. *Source*: Barbara K. Jones.

A more recent espouser of this agrarian ideal is poet and essayist Wendell Berry, a man who considers the presence of the family farm on the American landscape as a way to maintain society's oneness with nature.[27] To Berry, a farm family's survival is premised on its ability to unify diverse economic, cultural, and environmental forces. From this need, the commonality between nature's needs and the family's needs dictates an ecologically sound use of the rural environment. Throughout the first half of the twentieth century, this need might have seemed reasonable, but with the rise of the environmental movement and corporate farming, along with the numerous examples of agricultural pollution and habitat loss within and beyond the farmgate, and the change in how many Americans see nature, the metaphor "farming as the agrarian ideal" is not necessarily a meaningful way to encourage a new human-wildlife relationship.

According to Paul Thompson in his work on agriculture and environmental ethics, the environmental problems facing many farmers and ranchers today are the result of two failures. The first is the failure by environmentalists and philosophers to appreciate that agriculture is a production activity. Because production is the intentional transformation of something with less value to something more valued, it has not meshed well with the environmentalists' view that the intrinsic value of natural areas is in their untransformed state. The second is that farmers have not developed and defended an environmental ethic suitable for the post-Rachel Carson era. Instead, they have worked to preserve the status quo. In many cases, farmers have even aligned themselves with organizations that do not share their long-run interest in the productivity of the land rather than with environmentalists who share what should be the farmer's "commitment to permanence and sustainability."[28] Although farmers strive to live the "natural lifestyle" in that the "proximity to nature is highly valued," they typically describe themselves as businesspeople.[29] Such a description, as well as a business worldview, contributes to the general feeling that many farmers are no longer trusted stewards of the land.

As reliance on agricultural chemicals increased and small-scale cattle production that spreads natural fertilizer decreased, the farming and ranching practice of straightening streams that altered nutrient flows and dried out land became commonplace. The nutrients that used to cycle within the farm from the grass to the cow and back again are now chemical fertilizers that get leached into the streams and flow beyond the farm gate to oceans and bays. The result of nutrient dumping from both farmers and homeowners are algal blooms that can lead to eutrophication. In 2021, evidence of this was clearly seen with the disappearance of seagrass, a primary food source for manatees, in the Indian River Lagoon ecosystem off the east coast of Florida. The loss of food led to a massive manatee die-off. By November 2021, over 1,000 manatees had died from malnutrition and cold. In part, by failing to contain excess nutrients on their land, the agrarian philosophy defined by "ranching as a way of life" or "farming as the agrarian ideal" had failed the manatees. To resonate more broadly, members of our nation's agricultural communities need to see land stewardship as a process in tune with both their own self-interest and as a public good.

COEXISTENCE WITH NATURE

As a conservation ethic evolved in the United States, states developed hunting regulations and rules, while at the federal and state levels, wildlife and habitat protection acts were passed to encourage and enforce a relationship with nature that attempted to limit our human excesses. Yet tragedies of the commons persist in part because there is no shared conservation ethic that exists across the country. Although conservation as we know it today would be unfamiliar to Indigenous people pre-European settlement, hunter-gatherer groups did recognize family-held hunting territories, and Native American agrarians relied on a form of regulated commons for gardens and other shared resource spaces.[30] Stories and mythologies spoke of how, with the passenger pigeon, the Seneca believed the birds "furnished a tribute to mankind, because their Maker had selected the wild pigeon for this important duty," so as a "gift to mankind, the receipt of this largesse carried with it responsibilities: the pigeons could only be harvested at the proper time and in the proper way."[31] The proper way of managing the resource ensured that adult pigeons could breed again for the next generation of birds. Their stories forged a sustainable relationship with the pigeon. In contrast to the language of a "gift to mankind," according to Joel Greenberg in his work on these birds, as Europeans settled the land, the passenger pigeons' habit of traveling in massive numbers encouraged farmers to equate their presence to a "plague of locusts." Between their large numbers and destructive habits to agricultural fields, a pest narrative along with the realization that the birds were a cheap and easily obtainable source of food led to a tragedy that ended with their extinction.[32]

If we look at this relationship exclusively through the lens of dominion over nature, we miss that there is a tool that forces coexistence to occur. That tool is the Endangered Species Act (ESA), where once a species is listed, there are mechanisms and rules put in place that can protect both the species and its habitat. Yet forcing coexistence with species as an alternative storyline to dominion is not broadly welcomed. A recent conversation with a beaver advocate is a reminder of just how alienating the coexistence umbrella can be to those who should be on board but aren't. According to Dr. Suzanne Fouty, a retired Forest Service Hydrologist/Soils Specialist, after an ongoing drought and major fire season in Oregon, this should have been the moment to end beaver trapping and hunting as a recreational activity on federal land in Oregon. Yet even though beavers are a critical keystone species and scientists recognize the important role they play in mitigating the effects of climate change, a small number of trapper advocates prevailed to keep trapping in place. The failure to win this is one that Aaron Hall, a beaver specialist for Defenders of Wildlife, recognizes when he told me scientists don't necessarily have expertise in communication and outreach, so coexistence has become a "loaded word." Hall believes coexistence has been poorly defined, and knowing what it means and what it doesn't are not questions that have been clearly answered. The space this absence provides has let ideas about what coexistence means evolve outside of its intent. Once misinformation about coexistence takes root in a community, it becomes not only part of the community's resistance to outsiders but also embedded in their shared point of view.

Since coexistence is critical to our survival, its message requires a better and more inclusive story. One endeavor that is helping shift stories of what coexistence means is the success of wildlife corridors. Wildlife corridors critical to animal migrations are being identified and enhanced through the creation of wildlife highway overpasses, culverts, bridges, and better fences across the country from Florida to Wyoming. The success of these wildlife corridors is based on the role these crossings play in the reduction of animal and human injury or death. In Trappers Point, a critical wildlife crossing area in Wyoming, there were on average eighty-five wildlife-vehicle collisions each year. After building fences, overpasses, and underpasses, wildlife-related collisions dropped by 80 percent and the percentage of pronghorn and mule deer successfully moving through the area increased significantly. Their success turned those initially opposed to wildlife crossings into advocates for more. According to Matthew Kauffman, a wildlife researcher at the USGS Wyoming Cooperative Fish and Wildlife Research Unit, "Clearly, this is a problem we know how to fix," and it's one that cuts across stakeholder groups, from transportation agencies to Tribes to conservation organizations to hunters. Not only can properly designed wildlife crossings nearly eliminate collisions at ecologically critical points in the landscape, he said, "they [also] promise to reconnect fragmented habitat" and maintain the American West's own Serengeti with its great migration of ungulate herds.[33]

Elk feeding grounds are another attempt at what some might consider to be a form of coexistence. The feeding grounds were put in place in Jackson, Wyoming after a severe winter in 1908 when many elk died of starvation. People in the Jackson area were concerned by the large numbers of dying elk, so in 1909 private individuals began feeding them. In 1912, the federal government gave the Wyoming Game and Fish Department (WGFD) $5,000 to feed elk, and by 1929 there were three feeding grounds in Wyoming to offset future starvation events. The Wyoming legislature also passed laws to protect landowners from damage done by game animals like elk, so that by continuing to feed elk during the winter, the conflicts that could result from hungry animals intruding onto private land were minimized. Although the cost of feeding the elk is significant ($1.6 million in 2020, with $163,118 coming from special elk hunting permits), feeding elk has also maintained a robust game population for hunters in western Wyoming.

At an elk management meeting I attended in Cheyenne, Wyoming, the public was invited to discuss elk management and what to do about the elk feeding grounds. A problem frequently expressed by those in attendance was that the feeding grounds have encouraged an increased tolerance for unsustainably high numbers of elk, and those high numbers have over time spurred the transmission of disease within the herds. Diseases like brucellosis (which causes spontaneous abortions with first calves), scabies (a parasite that can crust over parts of the bull elk's neck, trunk, and upper legs), and chronic wasting disease or CWD (fatal infections from prions passed from animal to animal or from the environment to an animal) can readily spread when animals are concentrated together in large numbers. The recent discovery of CWD at the feeding grounds has added to the controversy over maintaining these sites. Even with the presence of CWD and the fact that feeding the elk allows their numbers to exceed the land's winter carrying capacity, many would argue there are reasons to maintain them. To ranchers, it is a boundary issue about keeping elk away from their winter hay, reducing transmission of disease to their cattle, and minimizing damage to their fences. To hunters, it is a consumption and trophy issue where a large enough herd must be maintained to satisfy their needs, while to tourists and wildlife viewers, seeing large herds of elk on the land is important to their enjoyment. Moreover, the revenue generated by the winter horse-drawn sleigh and wagon rides through the National Elk Refuge feeding grounds is beneficial to the local economy (see figure 2.4). Like the wildlife corridors, the feeding grounds appeal to a wide constituency so framing them as a tool of coexistence can potentially increase what is possible for a broader acceptance of coexistence as a lifestyle choice.

To become a shared way of relating to nature and wildlife, coexistence must mitigate boundary issues and points of view long embedded in people's imaginations. As coexistence broadens its reach, it becomes possible for constituents to code-switch and reimagine traditional stories and deeply embedded ideas. For one-third generation ranching family, living with grizzlies and wolves on their land taught them how to coexist. They recognized that the long-established way of running their ranch relied on a win-lose approach, where grizzlies and wolves had to lose, and they wanted a win-win. By adapting to a human-managed landscape and accepting new food sources, the grizzly bears showed this family of ranchers how flexible they can be, and that willingness became their ultimate teacher of change. They saw a win-win coming through the lens of "if grizzly bears can be flexible, then so can we."

Figure 2.4 National Elk Refuge Wagon Ride into the Refuge. *Source*: Barbara K. Jones.

WHAT WILL IT TAKE?

When a coyote entered an established residential area in Vancouver, Canada, loping down the road in broad daylight, environmental historian Graeme Wynn asked, "What was *it* doing swaggering—not even skulking, but strutting as though *it* had every right to be there—through human space in broad daylight? This was haughty behavior indeed in a creature that, for all its dog-like appearance, was considered a bother, a nuisance, a pest, and a trespasser."[34] This question posed by Wynn suggests that by crossing into the human-controlled sphere of a residential neighborhood, the wild coyote became a domestic nuisance. The use of that nuisance metaphor shifts our perception of the coyote from a wild animal free to roam to a pest that has become too emboldened for its own good.[35]

For those who see people living outside of nature on anthropogenic landscapes, there exists an imaginary border that separates "highly controlled" human spaces from "wild, uncontrollable nature."[36] To them, when wild nature border crossings occur and become disruptive, it is because the violator has failed to behave appropriately in a human-defined world. A colony of beavers that enters the domain of the rancher and floods the rancher's land or a wolf that stresses elk, making them more difficult to hunt, has in each instance crossed that "regulated" border created in our imaginations. By violating this space, those animals have defied the rules of "orderly civilization" and brought the unmanageable wild into our midst. Assigning negative human attributes to the offending species through stories and metaphors is part of a long-standing response to this defiance of our rules. Hence, the beaver became a pest, the wolf vermin, the bison a nuisance, and even the sea otter a thief. For coexistence to reflect our interconnectedness across a shared landscape, it becomes essential to reimagine these outdated wildlife descriptions so that our invented borders become less regulated and more welcoming to wild nature.

To achieve this, we need to think about the metaphors and stories that reinforce who can and should occupy our domestic landscapes. When we think about coyotes in New York City's Central Park or foxes in a suburban driveway, we are thinking about animals as outsiders because we think they don't belong in these places; instead, they belong somewhere else. Someplace wild. The problem with this solution is that the ideas of wild and wilderness are social constructs based on the notion that wilderness is a space untrammeled by man and wild is untamed. In both instances, they parallel a spatial understanding that is framed by what it means to live in highly managed spaces like cities or towns versus being in a setting that Aldo Leopold described as a distance you can travel in a two-week horse pack trip without crossing a single road. Each of these fails to capture how humans and wildlife are embedded in a shared nature where our space overlaps theirs within an interconnected web broad enough to welcome us all.

NOTES

1. Clark 7.
2. Olwig 27.
3. Brinkley 224.
4. Brinkley 229.
5. Brinkley 229.
6. Breining 26.
7. Brinkley 287.
8. Herrick 636.
9. Jones 2019: 198.
10. Greenberg 12.
11. Greenberg 109.
12. Organ, J. F. 19.
13. Ibid. 15.
14. McNally Oct 8, 2021.
15. Koshmrl Jan. 29, 2025
16. Koshmrl 10A.
17. Peterson 37.
18. *Save the Nautilus.*
19. Chaney 172.
20. Chaney 189.
21. Ibid. 189.
22. Phillips. WYIdlife Tourism for Tomorrow.
23. Jefferson 1984: 818.
24. Browne 1992; Berry 1996.
25. Browne et al. 1992: 10.
26. Browne et al. 1992: 10.
27. Berry 1996, 1981.
28. Thompson 1995: 15.
29. Thompson 1995: 14.
30. Krech III, Ostrom 1990.
31. Greenberg 43.
32. Greenberg 74.
33. Holland. June 10, 2020.
34. Loo XI.
35. Jones, April 11, 2022.
36. Soniak 11/14/16.

Chapter 3

American Bison

Why Is It Our National Mammal?

HOW DID WE GET TO NATIONAL MAMMAL STATUS?

With the passage of the National Bison Legacy Act (NBLA) on April 28, 2016, the American bison (*Bison bison)* was rendered a symbol of "unity, resilience, and healthy landscapes and communities" and selected as the United States' national mammal. Something extremely rare occurred when senators and representatives of both parties came together to unanimously pass this act. Five months after its passage, in the *Huffington Post* article "Bison Have Been Bringing Americans Together for Centuries," John F. Cavelli, vice president for international policy at the Wildlife Conservation Society, wrote how bison make finding common ground possible. For many, the NBLA provided the language to make such a claim: it identified bison as a historical symbol of the United States, and by recognizing the more than seventy tribes participating in the Intertribal Buffalo Council and the numerous tribal members involved in bison restoration on tribal land, it acknowledged the bison's long ties to the economic and spiritual lives of many American Indian tribes. It also accepted the important role bison play in maintaining the integrity of plains ecosystems. While the act made the bison our national mammal, it did not change any existing federal regulation or policy regarding the animal; it merely recognized that the shaggy bison with its red dog calves was woven into our cultural identity.

Place names that stretch from Buffalo, New York, to Buffalo, Wyoming, the presence of the American bison or culturally familiar buffalo on state quarters and flags, and the existence of the World's Largest Buffalo Monument in North Dakota suggest a powerful bison imagery exists within the American consciousness. By the time the first European settlers and hunters entered the western grasslands, there were so many bison their presence

Figure 3.1 Bison in Paynes Prairie State Park, Florida. *Source*: Barbara K. Jones.

defined the landscape. Their numbers were so great they naturally expanded eastward, with documented evidence of bison as far southeast as central Florida. In the eighteenth century, William Bartram, a renowned Florida explorer, wrote in his travel journal, "The buffalo . . . once so numerous is not at this day to be seen in this part of the country."[1] Yet today, hundreds of years after the last bison was spotted in Florida, as part of the Real Florida initiative by the Florida Park Service, Paynes Prairie State Park near Gainesville has become home to fifty genetically pure bison. According to Donald Forgione, the park's manager, of all the horses, alligators, and bison in the park, the bison are the "icing on the cake." Although the bison are constrained behind a fence, they roam across 22,000 acres with little human contact (see figure 3.1).

For the grassland ecosystems of the West, the bison story is different and sadly more defining of a shared American story. The grasslands that once supported millions of bison became home to a mere handful. Those that remained found refuge in places like Yellowstone National Park (YNP) and Charles Goodnight's ranch in Texas. The bison that had played a critical role in Native American survival and identity were shot by Europeans for sustenance or for those seeking their fortune. This indiscriminate killing of bison not only hastened the removal of Native Americans living on the Plains but also helped make room for the railroad and an influx of settlers. The prairie with its unique ecosystem that benefited from the large nomadic herds of bison and their stamping, wallowing, and mosaic-style grazing quickly became a land of farmers and plows. A market economy that didn't appreciate the finite nature of a species and a food-production system that relied on the plow had cultural, ecological, and economic consequences still felt today.

From this travesty, a conservation ethic rooted in saving the bison was born. In 1905, there were approximately one-hundred bison in public herds. The majority were in YNP, with five on live display at the Smithsonian in Washington, DC. In that same year, to save bison from extinction and raise public awareness about these unique animals, private citizens as well as William Hornaday, the first director of New York Zoological Park (Bronx Zoo), and President Theodore Roosevelt founded the American Bison Society (ABS). A few years later, the ABS moved to establish a free-ranging herd. In 1907, Hornaday, who had been raising genetically pure bison at the zoo, sent fifteen to Wichita National Forest Preserve in Oklahoma, the oldest managed wildlife area in the United States Fish and Wildlife Service system, and, in 1913, thirteen more to Wind Cave National Game Reserve. Today, Wind Cave National Park in South Dakota and Wichita Mountains National Wildlife Refuge in Oklahoma maintain parts of our national herd. By 1910, the ABS census estimated there were 1,076 bison in Canada and another 1,032 in the United States with an additional 151 in public herds. To accommodate the increasing number of bison, President Roosevelt created the National Bison Range on land taken from the Flathead Indian Reservation. In 1935, with total bison numbers in the United States at around 22,000, the ABS voted itself out of existence, thus ending their role in raising public awareness about the bison and preventing its extinction. For the ABS, a population that once numbered in the tens of millions was seen as recovered with only 22,000 animals (see figure 3.2).

According to Shana Drimal, Senior Wildlife Conservation Associate of the Greater Yellowstone Coalition (GYC), bison are the charismatic face of the plains, and while evidence from her work suggests they are the most popular megafauna, today, other than those found in YNP, most bison in the United States are considered livestock. Unlike the wisent, or European bison, with its thousands of years of recorded history, the American bison has had only about five centuries worth. Tom McHugh, a buffalo historian, acknowledged that even though the time for bison storytelling has been relatively short, writers have made up for it in frequency, making bison the North American continent's literary mammal superstar. This short-written history that began early in the sixteenth century established a specific way of seeing bison that relied heavily on the language of bison as livestock. Hornaday commented in his book *The Extermination of the American Bison* how one early Spanish explorer wrote the "greatest Rarity was the Mexican bull . . . It had crooked Shoulders, with a Bunch on its back like a camel; its Flanks dry, its Tail

Figure 3.2 Bison Behind the Smithsonian Castle on the Mall in Washington, DC. Taken between 1886 and 1889. *Source*: Photographer unknown. Smithsonian Institution Archives. Image #MAH-8008A.

large, and its Neck covered with Hair like a Lion" as the "first seen buffalo described."[2] McHugh noted how one Spanish explorer called bison "oxen" with "little hornes like the Moresche [Moroccan] cattle,"[3] while others simply called them obstinate cattle. Since wild bison are difficult to manage or fence, nearly impossible to herd, and not well-suited to serve as beasts of burden, maintaining a "bison as cattle" or livestock metaphor made it easier to substitute one cattle type for another. The sheer number of bison diminished their economic value while the apparent ease with which hunters and frontiersmen killed them made it simple for early Western settlers to pass down what Hornaday called a "contemptuous" opinion of these animals. By the 1870s, the bison that, at an earlier time, had formed "one robe" across the country was almost entirely replaced by more settlers and their more "valuable cattle."[4] The livestock metaphor along with a low opinion of bison as a nuisance species had so effectively maintained the "bison is not a wild animal" myth that many of today's cattle ranchers still rely on some version of that to insist bison remain in captive herds.

As captive herds subject to anthropogenic forces, bison are less likely to compete with other grazers, interact with wild predators, or influence an ecosystem through their actions. Because ranchers around Yellowstone fear bison will transmit brucellosis to their cattle, bison cannot move freely over the landscape as elk or deer can. Although elk are also carriers of the disease, it is only when bison leave YNP to find food that they once again can be, according to Rick Lamplugh, author and bison activist, "slaughtered," but this time by not only bison hunters but park officials as well. He described how, when bison leave Yellowstone, they travel the "final four miles [of their traditional migration route north] through Roosevelt Arch [YNP's north entrance] to Beattie Gulch," where they meet their end "trapped" for a "canned hunt" or, if they are lucky, are quarantined for transport to tribal lands as livestock. A visit to the gulch after the winter 2023 hunt ended introduced me to the extent of the killing. To get a better view of Beattie Gulch, I walked up a short trail that ran alongside YNP's northern boundary only to discover an intact bison carcass lying on the park's side of the boundary. The bison had most likely been shot in Beattie Gulch but then died in YNP outside the shooter's reach. To avoid an unnecessary confrontation with any grizzly bears that might have been in the brush around the carcass, I turned from the trail and walked back to the road that ran along the edge of Beattie Gulch. The presence of ravens and magpies directed my attention to the many bison skeletons littered across the landscape. The carcasses were left for scavenging by wolves, bears, and ravens, just as the white settlers had done in the killing fields of the plains one-hundred and fifty years earlier. It was disturbing to see so many carcasses left to rot, but the stories told around Gardiner, Montana, about how the bison were shot practically in front of people's homes and that gut piles were disposed of around town made the bison hunt a euphemism for something significantly darker. For those working to return wild bison to the prairie, concern about brucellosis is just another effort used to hide the age-old argument of bison competing with domestic cattle for limited grass (see figure 3.3).

The Intertribal Buffalo Council was established in 1992 to transfer bison from national park lands to tribal lands. It consists of seventy tribes in nineteen states that maintain a collective herd of over 20,000 bison spread across reservation lands that total thirty-two million acres. Of that land, one million acres are home to restored bison herds that include highly managed small herds of a few hundred or so to free-range large herds of a thousand or more animals. Currently, the Fort Peck Assiniboine and Sioux Tribes in Montana maintain both commercial and conservation or cultural herds. To initially establish these conservation herds, bison needed to be transferred from YNP to their tribal lands, but because of rancher opposition to the transfer, tribal groups spent years in legal battles with the state of Montana. In the publication *Yellowstone Bison: Conserving an American Icon in Modern Society*, Daniel Wenk, then superintendent of YNP, described an interesting example of just how serious the state of Montana was about keeping bison in the park. Bison moving into the Gardiner basin of Montana (an area north of the park) during the harsh winter of 2010–2011 precipitated state concern about wild

Figure 3.3 Bison Carcasses Left in Beattie Gulch after the Bison Hunt, May 2023. *Source*: Barbara K. Jones.

bison living outside the park. In February 2011, two weeks after his arrival at the park, Wenk was invited to Governor Brian Schweitzer's office to discuss bison management. One idea proposed by the governor was to hunt bison within the national park. Wenk rejected the suggestion because it was not supported by law, regulation, and policy for either the National Park Service (NPS) or Yellowstone. In addition, Wenk believed the park could ecologically support the bison and that there were appropriate bison management strategies already in place. The governor's response was to issue an Executive Order that banned the shipment of bison on state highways. To Wenk, this was his first lesson in the politics of bison management.[5] Ultimately, though, the Fort Peck Tribes' powerful arguments that tied bison to their culture and land won their case against the state of Montana, and today bison are being transported across Montana roads to more and more tribal lands.

This enthusiasm for returning bison to the prairie reflects a general change in our relationship with wild nature. With fears over habitat decline, climate change (the prairie is exceptional at sequestering carbon), and species extinctions, listing the bison as "ecologically extinct"[6] has encouraged a new wave of bison conservation enthusiasts. It also led, in 2006, to the re-establishment of the ABS by the Wildlife Conservation Society. In Montana, this enthusiasm is tempered by ranchers concerned over American Prairie's, a nonprofit formed to manage Montana's unique northern grassland ecosystem, goal of buying and then returning ranchland to native grassland habitat. To make their point, the region is decorated with signs that read, "Save the Cowboy:

STOP American Prairie Reserve." Yet the ranchers' usual persuasiveness is being met by tribal groups who, by bringing bison back, have made themselves a more visible and potent force in the West. As part of a larger reconciliation process, the National Bison Range land taken by the federal government in 1908 was recently returned to the Confederated Salish and Kootenai Tribes. This call to action has highlighted the role disparate groups can play in restoring bison to landscapes from the Intermountain West to the depopulated central plains (see figure 3.4).

As the bipartisan NBLA suggests, returning bison to the prairie has the potential to satisfy a broad constituency. In "Make Way for Buffalo," journalist Nicholas Kristof wrote in 2003 that "Th[e] forlorn farm town [of] Rawson [a now unincorporated farm community in North Dakota], population 6 [2000 census]—is a fine place to contemplate the boldest idea in America today: rescuing the rural Great Plains by returning much of it to a vast 'Buffalo Commons.'"[7] This idea had already been tested before when in 1919 bison were brought to the newly created Custer State Park in South Dakota for the sole purpose of generating revenue from tourism. Park officials there placed such hope in the bison that they vertically integrated them into many aspects of the tourist experience, where tourists could see not only living bison on the prairie but also view their slaughter and "taste the Old West" as well. Today, bison could become the attraction that brings vanishing rural communities across the prairie much-needed economic renewal. In the September 2020

Figure 3.4 A Save the Cowboy Campaign Sign on a Ranch Gate in Northeastern Montana. *Source*: Barbara K. Jones.

issue of the *National Review*, the journalist Santi Ruiz wrote how the online conservative "megafauna nationalists" also dream of rewilding America with bison. To Ruiz, a partnership that includes support from "unexpected places" could help build a powerful coalition pushing for the bison's return to large swaths of public land. Western tribal groups, bison conservationists, East Coast *New York Times* readers, and online conservative "megafauna nationalists" (who Ruiz believes have a "visceral hunger for moonshot projects"[8]) working together with affected communities might be the right combination of unexpected partners.

Yet for all this bison love, there are still too few bison in conservation herds. According to data presented by the GYC, recent counts estimate the number of bison in the United States today to be 450,000, with five percent of those bison, or about 22,500 animals, in conservation herds and the remaining existing as confined domestic livestock.[9] As livestock, the bison's wild megafauna essence is reduced to that of a non-native Hereford cow, while as wildlife, we really don't know our cultural carrying capacity for bison freely moving across the plains. Although passing the NBLA was a critical step in redefining our relationship with this charismatic wildlife icon, returning bison to our shared landscape might take the unity of purpose forged through some unlikely American partnerships that have a shared way of seeing coexistence.[10]

THE MYTH OF THE AGRARIAN IDEAL

To understand the role of myth is to understand how it reinforces what many people see as their cultural ideal. Ideal culture describes those behaviors that people say they do or want to be true, as opposed to real culture, which describes what people actually do. Myth is more about ways of life or convenience than it is about fact, in that it promotes what many people want to be real, even if that reality is no longer useful or relevant. Beliefs and values that are tied up in these myths organize and inspire people's responses to social problems and since they help form an individual's worldview, any challenge can not only trigger significant pushback but can also feed into larger cultural issues. The ways in which the scientific evidence for human-induced climate change has divided people who share the same landscape into such angry opponents show us just how powerful myths can be. Since our beliefs and values help form our identity, challenges to them can become highly personal and hotly contested.

In 1785, prior to the establishment of the United States as a nation, there were constitutional concerns over how to reconcile private interests with the public good. The idea that individuals tended to shirk public responsibility led to a significant debate between Thomas Jefferson and the Federalist Alexander Hamilton over whether power should be centralized or widely dispersed. The Federalists, who believed some people were better at recognizing the public good, looked to large landholders as the most valuable allies to a stable state and as such would be more reliable citizens, while Thomas Jefferson argued that since the common farmer, not the aristocrat, was the primary occupant of the land in America, it was the yeoman farmer's self-interest and the common good, defined then as "the stability and fiscal responsibility of the society," that overlapped. To Jefferson, this convergence of public and private interests was "a mark of citizenship" that, although society would change over time, once this blend of interests became established as part of America's democratic process, new ways of serving the public good would emerge.[11]

To Wendell Berry, the maintenance of social ties in rural America and the necessity of good land stewardship require the virtues of "honesty, integrity, and charity"[12] that are also critical to democracy. This inclusion of stewardship makes room for a modern environmental ethic and reminds us of Jefferson's recognition that although society would change, a strong foundation in the democratic process would continue and adapt as new ways of seeing the "public good" became identified and valued.

The connection between Protestantism and hard work identified by sociologist Max Weber reminds us of this agrarian value system. By tying production, fertility, and fecundity with moral virtue, it became possible to see farmers as virtuous people. Their efforts resulted in increased production, which was both a reward and a sign to others that the farmer was hardworking and possessed moral virtue. Protestants with Calvinist roots believed that God's chosen would live moral lives and His chosen people would be revealed through visible signs. One sign was fertility or fecundity, and so those who not only produced children but animals or crops were evidence that God's grace had smiled upon them.[13] Growing the biggest pumpkin or raising the largest number of calves became integral to a farmer's identity and was rewarded with prizes and recognition in community events like county fairs and rural grange contests. This belief system understood implicitly that these virtues were highly valued outcomes, and by not clearly expressing them, they became almost mythical in their effect. Over time, though, seeing fecundity as a virtue worked against the public good because as technology improved, productivity increased to exceed America's food needs. Decreased incomes from overproduction and the depletion of soil and grass that should have been market signals were masked by America's farm policies and the values deeply held by many farmers and ranchers. From that, some agrarian producers became less inclined to balance their self-interest with the public good. Today's environmental ethos conflicts with a farmer's belief and value system that sees overproduction as a virtue because it fails to consider the environmental costs associated with that extra production. The true value of farm output needs to include not only the net income from a farmer's saleable products but also the costs associated with the externalities of pollution and the loss of wildlife and wildlife habitat that can occur when the public good is discounted or overlooked because production is prioritized.

Thomas Jefferson and later enlightenment thinkers like Ralph Waldo Emerson and Henry David Thoreau saw farmers as the embodiment of moral and political ideals that should be applied to all citizens, yet today those same arguments for farming as the agrarian ideal have been used as reasons for exempting farmers from those criteria. By the 1970s, the efficient allocation of society's resources was the explanation for the decline of America's heavy industry, but somehow farming was exempt from addressing those inefficiencies.[14] This exemption from the rules encouraged some urban and suburbanites to see farmers as "glorified welfare recipients or as willful polluters, rather than as paragons of virtue,"[15] while their failure or refusal to adapt to changing notions of what constituted the "public good" defied the agrarian values of hard work, honesty, integrity, and charity. Just like other myths, the agrarian ideal has not withstood the test of time, yet it persists as a metaphor for the exceptionalism of farmers. For a small group of powerful farmers and ranchers, that exceptionalism is no longer rooted in articulating agrarian ideals that balance self-interest with the public good, but instead on maintaining their way of life. From that, one can only wonder: do people today see ranching and farming communities as the principal source of moral inspiration and virtue?[16]

In twenty-first century America, the irony that an agrarian philosophy rooted in the importance of community and the public good is now be used to promote the interests of a few is disturbing. Since there are many farmers who welcome milkweed for butterflies and maintain vernal pools for birds, add solar panels to their homes and barns, set aside habitat for wildlife and wild nature, and maintain open spaces for our common benefit, it is not fair to say all farmers or ranchers operate without regard for the greater good. In Montana, there are ranchers who manage working cattle operations, yet they temper their activities to allow predator-prey relationships to play out. In North Dakota, a public opinion survey regarding mountain lions (cougars) in the state suggests that even though mountain lions can impose significant costs on ranchers through livestock predation, across all survey groups, the presence of mountain lions was met with overall approval. For this, the few then is a small minority of powerful ranchers and farmers who use their influence to maintain a myth that no longer promotes the importance of community and the public good. A reliance on the agrarian myth has allowed an influential few to define how public land is used and how wild nature is managed, as if ideas about land use were fixed in the distant past and their voices still speak for the majority.

A particularly egregious example of this betrayal of principles of honesty, integrity, and charity and the public good is playing out with bison. In most places, bison are seen as livestock and are managed under the auspices of the United States Department of Agriculture (USDA) or a state's department of livestock. Bison are typically not allowed to roam as wild animals because ranchers fear bison will transmit brucellosis to their herds. There is no evidence of this transmission from any wild Yellowstone bison, yet Montana's zero tolerance for brucellosis in cattle has successfully been tied to strict rules and restrictions on bison. According to scientists, the risk of brucellosis transmission from Yellowstone bison is insignificant (less than 1 percent) compared to elk (more than 99 percent of the total risk) because elk overlap in more areas with cattle and are more socially tolerated by those working with cattle.[17] To reduce the costly economic impact on ranchers in nonaffected areas, the USDA's Animal and Plant Health Inspection Agency (APHIS) established a rule in 2014 that required the creation of a brucellosis management plan for any state that had brucellosis-infected wildlife. While it is important not to minimize the economic impact of brucellosis on the sale of cattle, because when Montana lost its brucellosis Class-Free status in 2008, it cost Montana ranchers $11.5 million due to herd quarantines and destruction, as well as travel restrictions, it is curious that the species of concern has consistently been bison rather than elk. To Shana Drimal, brucellosis is a "smokescreen" issue as to why bison can't leave YNP. She believes ranchers worry that cattle will be displaced by bison on public land, yet as soon as bison leave YNP, they are managed by the Interagency Bison Management Plan (IBMP). They are not free to move along their ancient migration corridors or seek food outside the park. By denying bison access to these migration paths, information about where to find food can disappear from the herd's memory; moreover, if the bison that leave the park are culled, that knowledge can be lost to YNP's unique founder population forever. To transition bison into wild spaces and to keep their evolutionary processes intact, the GYC wants to manage national forests as corridors for movement to improve access to tolerance zones. Since Yellowstone bison have brucellosis, they are not free to roam; instead, they are hazed back into the park, or they are removed from any area that falls outside the bison's designated surveillance area. This area, which is primarily in the GYE and includes tolerance zones north and west of YNP, is a political boundary because it is "illegal" for bison to be outside that designated space.

A challenge for those working to expand the footprint of wild bison is the way bison migrate out of the park in the winter. Historically, the park's central bison herd had migrated out of the west side of the park, but now more herds are migrating north into Gardiner and Beattie Gulch. To Jason Baldes, Tribal Buffalo Program Manager for the National Wildlife Federation (NWF) and a member of the Eastern Shoshoni Tribe, the Shoshoni Tribe is not an advocate for the hunt in Beattie Gulch. Their preference is a fair and honorable hunt, not a slaughter line. For the GYC, along with Defender's of Wildlife, and tribal groups like the Eastern Shoshonis, improving bison access to tolerance zones and expanding the zone's size are critical to bison recovery because bison are now locked into Beattie Gulch. There is no decent corridor into the Upper Gallatin area (national forest land) because of dude ranches and other competing uses, while the other access corridor through Tom Miner Basin (also north of Gardiner) is legally off-limits to bison because it is out of the designated surveillance area.

The bison-brucellosis story that has kept bison contained behind fences or within the boundaries of YNP and a few other parks and refuges is an example of how stories and myths can maintain our invented or imagined boundaries. To Todd Wilkinson, co-founder of *Yellowstonian*, and founder of and writer for the *Mountain Journal,* a nonprofit public watchdog publication for the GYE, for society to expect a wild animal like the bison "to remain contained behind an invisible line drawn on a map that does not conform to the needs of the biological species defies not only logic but the laws of nature."[18] Yet the language we use to frame these species does exactly that. By learning about wild animal behavior, it becomes possible to reframe the language and shift a nuisance animal to something akin to a partner rather than a pest. This in part explains why elk that also carry brucellosis remain free of the brucellosis stigma, and even with the reintroduction of the wolf, are maintaining populations in Montana that in some regions are at 200 percent of the land's capacity. According to Drimal, there is strong support for elk, so they are not subjected to the same constraints as bison, and there is little political will for hazing elk back into the park or shooting them in a "canned" hunt. This support allows elk to move freely across boundaries. In Wyoming, Idaho, Utah, Oregon, and Washington elk are also fed in the winter at feeding grounds (Wyoming bison are occasionally fed at the National Elk Refuge as well), while bison that leave YNP in search of food in the winter get hunted, slaughtered, and, more recently, quarantined for return to reservations.[19] How then did bison get tarred with the brucellosis feather, yet elk remain free to roam and live (other than at the feeding grounds) in the condition of their evolution? Such a question reminds us of the power deeply rooted myths and stories about bison have had in how rules and decisions are made regarding their management.

THE BISON AS LIVESTOCK CONUNDRUM

To understand how people take a position on topics like domestic versus wild bison, we need to consider work done by University of Maryland Communications Professor J. E. Grunig's on the idea of "publics." Grunig sees a public as "a specialized group whose members have a reason to be interested in the activities and behaviors of organizations,"[20] and respond to the behavior or activity in varying degrees that depend on their feeling of control over the outcome. Since these publics or groups form when affected people are left out of the decision-making process, Grunig argues there is no such thing as a "general public" or a permanent public, but rather publics that assemble and disassemble as situations change course.[21] For bison, there are two competing publics: those who want bison to become more of a public trust animal and those who want bison to remain as livestock. Up until recently, thc agrarian ideal metaphor or myth has kept the ranchers and politicians in control of the bison's fate as livestock. Yet, by leaving bison supporters out of the decision-making process, that control has encouraged a diverse set of bison enthusiasts to work together to alter the bison's fate.

Keith Stamm and James E. Grunig[22] examined attitudes people had about the environment and conservation. They found that functional substitution and reversal of trends can help explain how people respond to conservation and other environmental issues. Functional substitution relies on the idea of fungibility, where one species can substitute for another, allowing the needs of the ecosystem or environment to be met. Reversal of trends offers a solution to resource scarcity when the resource is used, but its use is limited to reduce the resources downward spiral toward scarcity. Their research suggests that environmental publics tend to rely more heavily on the reversal of trends solution unless the environmental problem is perceived to be personal. Once it becomes personal, both strategies can be used, or the affected public resorts to what the researchers identified as hedging. Hedging, like the reliance on myth, allows for the formation of a fuzzy space that can be filled with distortions and social amplifications that serve to counter the power of the opposing publics' hedge or myth.

For bison, cattle replaced them on the landscape to the detriment of the grassland ecosystem. The replacement by cattle was initially considered to be a good swap as cattle were more profitable to raise and easier to manage, but they didn't live on the landscape like bison. Bison, with their powerful hooves and constant feeding, manipulated the grassland, which extended the growing season and created greener, more nutritious grass.[23] It took far too long to recognize that cattle and bison were not fungible species. The reversal of trends acknowledges that replacing some cattle with bison not only reduces the pressure on the grass resource, but with the return of the coevolutionary processes of bison and grass working together, the scarcity of both bison and grass declines. The persistence of the "killing fields," as Todd Wilkinson described the winter hunt/slaughter of bison at Beattie Gulch, makes the bison issue extremely personal to many bison supporters. Once the issue shifts to the personal, all strategies are used, including the hedge. The hedge encourages bison enthusiasts and critics to operate in a somewhat broadly defined space, where, for example, carrying capacities are roughly defined (YNP bison biologists have offered park carrying capacity numbers that vary significantly) or negative outcomes are amplified. By giving enthusiasts wiggle room with carrying capacity and those opposed to wild bison the opportunity to forward negative outcomes, the hedge can potentially offer space for concessions on both sides.

The bison-as-livestock story has continued unchecked through much of the twentieth century. By 1962, Canadian bison were so much a part of the agricultural scene that Walter Dinsdale, minister of Northern Affairs and Natural Resources, could comfortably announce that "250,000 pounds of top-quality meat from Wood Buffalo National Park [will] go on sale from Jan. 17 . . . Everything from streamlined hindquarters to packaged buffalo burgers has been prepared to give jaded Christmas appetites a life."[24] In the 1960s, Wood Buffalo, Buffalo, and Elk Island National Parks in Canada had everything from chutes to abattoirs to accommodate their increasing trade in bison meat. The boundaries of Elk Island National Park were expanded twice to grow hay, oats, and green feed for the bison herd that had been fed there since their arrival in 1907. In response to the park's managers' worries that the feed requirement would only increase as the rising number of bison put pressure on the park's grasslands, in 1956 more parkland was put into feed production and then in 1957 an additional 600-acre farm was added to Elk Island National Park. The feed grown there fed the park's bison as well as animals at other parks. In 1967, when Canada was celebrating its centennial year, some of the country's wildlife parks had become stockyards. Not only did wardens work as bison cowboys, but the Canadian Wildlife Service maintained a reindeer ranching operation at Mackenzie Delta. This shift by the institutions that had been instrumental in the rescue and recovery of bison in the early 1900s diminished the bison's status as an iconic species of the North American conservation movement, a species that symbolized what many saw as the disappearing wild to something akin to "meat."[25] To support economic development in the Northern regions of Canada, Tina Loo, in her work on *States of Nature*, wrote how the federal government transformed the North's natural resources into a "national commons for the benefit of all Canadians."[26] This transformation of bison from a conservation success story into an agricultural food product was happening in the United States as well, making it challenging to reclaim bison as an example of the returning wild.

The bison's long absence in our imaginations made it more difficult to reintegrate them into wild landscapes. For most of the twentieth century, bison were recovering from near extinction, so they did not regularly wander out of YNP. Their absence on nonpark land led some to believe that bison should remain in the park and only leave when they exceeded the park's available grass. To scientists working with Yellowstone bison, this mindset has allowed wild bison to be treated as livestock when they move outside the park.[27] According to Rick Lamplugh, treating elk as trophies and bison as livestock creates an understanding that one is wild and free, and the other is no different than a sheep or goat. Yet to Shana Drimal, bison get a bad rap just by being a lot of things cattle are not. Cattle were bred through artificial selection to be calm, slow, tractable meat-makers that are relatively human-dependent, while bison through natural selection have evolved to be resilient, fast, independent, competitive, and wary.[28] Chamois Andersen of Defenders of Wildlife also finds the livestock designation frustrating because bison look wild, with their big heads and big hides, and to maintain their wild ways they need to be wild.

Those wild ways can make coexisting with them challenging. Bison are the largest North American land mammal, with the male weighing up to 2000 pounds and standing six feet tall; their eyes do not reflect light, so they are very difficult to see at night; they tend to move in large groups; and they can run up to thirty-five miles per hour, turn quickly, and can jump fences. If a car traveling sixty mph at night on Route 89 north of YNP hits a poorly visible, wild, free-roaming herd of bison, the consequences could be dire. This makes their presence on high-speed roads a significant public safety issue. That public hazard could extend to dude ranches or public lands as well, where both people and horses would have to navigate around extremely large, quick, and dangerous wild animals. The process of educating visitors is already a challenging one, where tourists visiting YNP have been known to "rescue" a bison calf or try to get too close for photos or to even "pet" one, but once bison are no longer considered livestock, the process of education becomes like that of any other wild animal. As more people travel into nature, the learning curve would begin to flatten.

Accepting there are challenges to maintaining wild herds of bison, the metaphor of bison as livestock is a dishonest one that has left the animal in an indefensible place. This categorization makes it logical to ask how can our national mammal that has the power and significance of the bald eagle can be seen as a domestic cow? Yet unless bison can find a way to expand their range on their own, which, under current rules in most places, would be very difficult, they will continue to be denied public trust status. As livestock, bison are not managed for all of us but instead for a select few who rely on the myth of the agrarian ideal to maintain their way of life. If the shift to bison as a public trust animal became a reality, the animal's wild presence could provide hunting opportunities, as well as generate income from bison tourism. Rather than a canned hunt in Beattie's Gulch or one arranged by outfitters on their private ranches, a tag hunt becomes possible if bison are treated as wild animals. According to Drimal, even though bison are more popular than any other megafauna, "old-time ranchers" are afraid of change, and so in areas west of YNP where bison could now be restored, ranchers and dude ranches are opposed, and they have a lot of power. Even "mom and pop" ranches are becoming "monopolies with big influence." To assist with bison restoration, groups like the GYC are working to help make human-bison coexistence possible through education and better fencing. By building bison exclusion fences and taking over retired grazing allotments and land leases, the hope is that social tolerance for bison will increase. But even though it is possible to add new sources of revenue from tours or hunting and provide barriers between wild bison and cattle, there are some ranchers who see accepting bison restoration as a betrayal of their values.

By not framing coexistence as an irreconcilable dichotomy, organizations like GYC, Montana Wild Bison Restoration Coalition, Defenders of Wildlife, and American Prairie in northeastern Montana are working to move wild bison into more people's consciousness. For the GYC, taking "baby steps" to establish relationships that will work for the long haul are essential, but without compromise, Drimal believes, "[they] aren't getting anywhere." This is particularly true with ranchers who see American Prairie as nothing but a group of "wealthy out-of-staters" who want to put wild animals back on the land. They are afraid American Prairie will acquire grazing rights from the Bureau of Land Management and then take down some of the existing fences to restore grassland and bison on not only public land but land that American Prairie purchased from willing sellers. To help return Montana's northern grasslands to a more optimal ecosystem, American Prairie hopes to take bison and shift them into a public trust wild species. As livestock in Montana, bison are currently subject to a per capita livestock fee, so to stymie efforts by groups like American Prairie, the Montana legislature passed a rule that any bison behind a fence would be charged a per capita fee and could never be wild. With the passage of that rule in 2021, American Prairie's model of maintaining some fences as they shifted domestic bison into wild bison could not occur. This was significant in Montana and could have been in other states as well because, with that rule, bison could only be wild in YNP and some small herds managed by the federal government, leaving all other herds as domestic animals, even those on tribal lands. Fortunately, legal action by Earthjustice helped negotiate an agreement that calls for American Prairie to test for brucellosis in a representative subset of its bison herd and share the information with the community. To build trust and common ground between the two groups, members of the ranching community have been invited to attend all handling events. According to Earthjustice, the plan demonstrates that large-scale rewilding projects are possible if you are willing to adapt to or overcome "significant cultural, economic, and legal barriers that make rewilding so complex on the ground."[29]

In Montana, the national mammal designation changed nothing for bison, so even though Montana is quick to use the bison as a symbol of the state on license plates, state coins, and schools, there has been very little tolerance for allowing bison on American Prairie land. The problem for those in favor of the bison's return to the wild is overcoming attitudes formed during their absence. Even for Native people, the connections with buffalo lost for over one-hundred years need to be retaught and reclaimed. For bison to return to the land, there are three pieces that need to come together: public trust, conservation, and culture. As a public trust animal, a marker of bison recovery can be made when their compensatory value can be augmented by hunting; to be huntable, a sustainable population must exist. To achieve the goal of a sustainable population, there are active public relations campaigns in place to elevate bison awareness that include bison "display herds," education, storytelling, and the American Prairie's efforts at reestablishing an American version of the Serengeti. To change the rancher's perception of bison as livestock will take new stories as well as a willingness to see coexistence as beneficial to their way of living. Efforts at improving bison access to bison tolerance zones north of YNP, maintaining bison exclusion fencing, expanding the bison conservation transfer program, and continuing to buy out grazing leases to minimize bison-cattle contact are efforts being made by a large coalition of bison advocacy groups. Even the revision of the twenty-three-year-old bison management plan is being rewritten with ranchers in mind (see figure 3.5).

To gather more of us under this coexistence umbrella, ranchers need to recognize that bison are not a threat to their cultural heritage. The feeling that bison compete with cattle for grass is one that ties into a social problem regarding ranching success and high productivity. High production rates might satisfy the rancher's self-interest and economic needs, but they can fail to serve the public good. For ranchers and farmers, Jefferson's ideal about blending the needs of the rancher and farmer with the public good should resonate with most producers of agricultural goods. It makes little sense for land conservation or stewardship to conflict with land uses as varied as cow-calf operations, soybean farms, and hunting, fishing, mining, or logging. All these activities ultimately benefit from maximizing our ecological role in ecosystem health without suppressing the wildlife populations that enhance the land's resilience. Several recovered mountaintop strip mining sites in Kentucky and Virginia are now lush meadows for reintroduced elk, while cow-calf operations typically maintain large sections of open land that can

Figure 3.5 Yellowstone National Park Bison Biologist Chris Geremia (right) Monitors the Bison Loading onto Trailers for Transport on January 12, 2023. The bison will then complete their quarantine process at Fort Peck Indian Reservation before being distributed to Tribal lands as well as other appropriate public lands. *Source*: Greater Yellowstone Coalition/Emmy Reed.

provide habitat for both the rancher's livestock and for resident wildlife. In both instances, if the land was not managed sustainably for the future, there would be fewer potential sites for elk recovery in Virginia, and poorer soil would mean less grass for ranchers' cattle.

To maximize the benefits of a sustainable land ethic, balancing the goals of production with those of sustainability could help public and private grassland recover, all while providing space for bison that, with their mosaic-style grazing and frequent wallowing, can help improve the overall carrying capacity of the land. Ranchers claim to be very in tune with the environment, and for most, managing the land in a sustainable way is a priority, making this an appealing message. To Scott Heidebrink, the bison restoration manager at American Prairie, "the ranchers all love big wide-open spaces. And they love wildlife," he admitted, and "that was probably the most surprising thing to me is how much we actually all have in common."[30] From a rancher's perspective, the environment might not always be uppermost in their minds because they see their way of life under threat from many fronts, where their children are abandoning ranching as a way of life, and there is a decline in recruitment to that lifestyle, in beef consumption, and for some, in family ties. When you add bison, wolves, grizzly bears, and even elk to the equation, the attack on their labor and subsequently their productive values become untenable. To one rancher, "Wolves are no good because they take my calves, while bison are no good because they compete for my grass." If coexistence is going to expand its reach, this is a worldview that is essential to understand.

By shifting the language of ranching and farming from that of a businessperson to one of "land caretaker or steward," an identity that balances production with the common good as the measure of success can make room for a shared coexistence story. By preferencing the words "land caretaker" or "land steward," some might think the social, ecological, and economic benefits of production-based models of land use are being minimized, yet the finite nature of our land resources suggests most land uses *must* emanate from a sustainable land ethic. Whether you are a farmer, rancher, hunter, outdoor enthusiast, or a "couch potato" who lives on cheese doodles and soda, maintaining the resilience and sustainability of the land should be of primary importance to every one of us. Curiously, early white settlers to Australia assumed the lush, green pastures they consistently found and felt would be perfect for grazing livestock were natural; yet those grasslands with the most fertile soils and fewest trees were the work of Aboriginal people who managed the land for productivity through the tenets of sustainability, abundance, convenience, and predictability.[31] Productive land uses do not have to be in conflict with the goals of conservation and land stewardship.

The need for a shared sustainable land ethic forces us to ask: since land is available, and bison do many things so much better than cattle, do we want cattle at the expense of bison monopolizing our nation's grasslands? If not, the ranchers' compact with the agrarian ideal requires them to service today's understanding of the public good as they work for their own self-interest. Since the behaviors of wild bison are critical to the health of grasslands, sharing the land with them can only enhance the grass necessary for cattle as well as any other species that shares that space. By having cattle and wild bison coexist on the land, there is hope that the agrarian ideal can become compatible with a twenty-first-century environmental ethic and the demands of the common good, as well as more appropriately honoring the bison as our national mammal.

NOTES

1. Bartram 62.
2. Hornaday 1.
3. McHugh 41.
4. Hornaday 29.
5. *Yellowstone Bison* X.
6. Defenders of Wildlife, Bison.

7. Kristof. Oct 29, 2003.

8. Ruiz. September 13, 2020.

9. *Yellowstone Bison: Restoring a North American Icon.* Park County Community Conservation Series. September 20, 2023.

10. Jones, March 26, 2021 (a significant part of this section was taken from this article).

11. Browne 6–8.

12. Browne 7.

13. Browne 51–52.

14. Browne et al. 11.

15. Browne et al. 15.

16. Browne et al. 15.

17. *Yellowstone Bison* 24–25.

18. Wilkerson Oct 2017.

19. According to Rick Lamplugh, at the Stephens Creek Capture Facility there are armed guards to make sure no bison escape transport. There are one-hundred hunters in a line that include 40–50 Native American hunters and 50 state hunters. Since it is an extremely small space to hunt, some Native Americans don't see it as fair chase, something they would prefer.

20. Grunig 2005: 778.

21. Grunig 2005.

22. Stamm and Grunig 1977; Grunig and Stamm 1979.

23. Earthjustice.org October 5, 2021.

24. Loo 121.

25. Loo 122.

26. Loo 122.

27. *Yellowstone Bison* 137–138.

28. Bailey 136.

29. Earthjustice.org Oct 5, 2021.

30. Earthjustice.org Oct 5, 2021.

31. Ashby 297.

Chapter 4

Caribou

When Being Called a Reindeer Must Really Matter

IS IT A CARIBOU OR A REINDEER?

The Eurasian reindeer (*Rangifer tarandus*) and its North American equivalent, the caribou (also called *Rangifer tarandus*), play a vital role in the survival of Arctic peoples in North America, Europe, and Asia. Although the caribou and reindeer are the same species, they have taken on vastly different cultural roles. According to Eric Chaline in his book *Fifty Animals that Changed the Course of History*, the partially domesticated reindeer in Europe and Asia served as the cow or horse[1] to people living above 66.5° north of the Equator. By providing food, clothing, and transportation, the reindeer's presence on the land helped Arctic people survive in a challenging environment. Although Chaline compares them to the cow or horse, reindeer were never fully domesticated, so their appearance and behavior have undergone fewer anthropogenic changes from processes of selective breeding or intense human control and management than domestic cows and horses. Today, there are Eurasian Arctic people whose economies are dependent almost entirely on reindeer, while the European Union's only recognized indigenous people, the Sámi, have relied on these semi-domesticated animals for over 2,000 years. Currently, Northern Europe is home to over one million wild reindeer and nearly three million semi-domesticated reindeer. Although climate change and habitat loss can challenge the semi-domesticated reindeer's survival, their semi-domesticated status protects them in some ways from the forces of extinction that can and have impacted wild populations of reindeer and the North American caribou that include boreal woodland caribou (*Rangifer tarandus caribou*), barren-ground caribou (*Rangifer tarandus groenlandicus*), Peary caribou (*Rangifer tarandus pearyi),* Porcupine caribou (*Rangifer tarandus granti*), and obviously the now extinct Queen Charlotte Islands caribou (*Rangifer tarandus dawsoni*).

The reindeer's adaptations to an Arctic lifestyle make them well-suited for survival in a climate of severe cold, as well as an abundance of snow and ice for a significant portion of the year. Their noses are designed to warm the air before it reaches their lungs, and the water vapor from exhaling gets retained within the animal's body. They have a double layer of insulating fur that not only keeps body heat trapped but also lessens the expenditure of critical energy. By expanding in the summer when the ground gets soft and contracting and hardening in the winter so they can break and dig through the ice and snow, their hooves are equally well-adapted. They feed on lichen, which is a symbiotic association of green or blue-green algae and fungus, while a shared adaptation for both male and female of this deer species is that both sexes have antlers. The males shed their antlers in early winter, but the females keep them until the spring, so they are better able to defend food supplies during their winter pregnancies.

Although the Genus Rangifer that includes caribou and reindeer is circumpolar, the "caribou" name is specific to North America and "reindeer" to Europe and Asia.[2] Wild caribou and wild reindeer are the same animal, where the difference occurs is when a reindeer becomes semi-domesticated. Hence, while some caribou and reindeer subspecies move, making massive migrations of thousands of miles as they travel between their breeding and winter grounds, for semi-domesticated reindeer, the processes of domestication have shifted them away from a life as truly wild animals that migrate to one that lives on smaller, more permanent grazing ranges. Since caribou were never domesticated and remain as wild today as they were thousands of years ago, they have typically remained lean and massive, with the quite small Peary caribou being an exception. Unlike the large body size of most caribou, wild Svalbard reindeer as well as semi-domesticated reindeer are shorter and stockier. Another difference between caribou and reindeer is that the reindeer's breeding season happens earlier in the year than the caribou's, making reindeer calves more susceptible to late winter predation from hungry bears. Climate change, predation, and human activities are forces that every wild reindeer and caribou must navigate (see figure 4.1).[3]

The social and economic importance of caribou and reindeer to people of the boreal forests and open tundra of the Arctic helped embed the native *Rangifer* species into their consciousness through myths and rituals. Over time, these myths became powerful stories that passed a way of seeing these animals down through the generations. Although reindeer have long since been extirpated in England, archaeological evidence tells us they once existed there before the last ice age and that overhunting led to their disappearance. Even without archaeological evidence, though, reindeer

Figure 4.1 A Limited Food Supply Keeps Svalbard Reindeer Small and Stocky. 2022.
Source: Barbara K. Jones.

stories, festivals, and rituals remind people in England today of the animal's importance to their ancestors. In Abbots Bromley, a village in Staffordshire, England, the community holds a yearly folk dance known as the Abbots Bromley Horn Dance. The horn dance includes dancers carrying six sets of one-thousand-year-old reindeer antlers as the center of the celebration. Since the "taking of horns" was a fertility symbol to local farmers, this dance maintains aspects of an ancient fertility dance and has persisted through tradition into the twenty-first century. The dance first performed at the Barthelmy Fair in August 1226 is today one of England's longest-standing medieval rituals.[4]

Sámi stories of Beivve, the goddess of the sun and mother of mankind, and the sacrifice of a white reindeer to her so that winter would end, along with the story of the brothers Attjis and Njavvis who, during a period of chaos, were murdered, leaving their wives to become the first reindeer herders, are critical to a Sámi understanding of the world. To the Inuit, caribou lore is closely tied to caribou hunting. One myth relates to the white caribou, where a woman looking for driftwood, bones, and caribou antlers got lost in the fog. She was rescued by a caribou hunter, but, during the hunter's absence, an evil shaman turned her into a white caribou. Since the caribou hunter's grandmother was also a powerful shaman, she gave her grandson the magic necessary to break the spell so he could turn the woman back into her human form. Today, Inuit caribou hunters protect the white caribou because they believe it could be an enchanted person.

A popular holiday reindeer story that captured the imaginations of people around the world is the poem by Clement Clarke Moore called "A Visit from St. Nicholas," with its most famous line 'Twas the night before Christmas.' As an explanation of what happens the night before Christmas, Moore wrote the story of St. Nick and his eight named reindeer:

> When what to my wondering eyes did appear,
> But a miniature sleigh and eight tiny reindeer,
> With a little old driver so lively and quick,
> I knew in a moment he must be St. Nick.
> More rapid than eagles his courses they came,
> And he whistled, and shouted, and called them by
> Name: "Now, *Dasher*! Now, *Dancer*! Now, *Prancer* and
> *Vixen*! On, *Comet*! On, *Cupid*! On, *Donner* and *Blitzen*!
> To the top of the porch! To the top of the wall!
> Now dash away! dash away! dash away all!"[5]

From this, the story of Santa Claus or St. Nick and his eight reindeer[6] became fixed in the hearts of families around the world. With radio and television programs, as well as children's books featuring Santa and his reindeer becoming classic Christmas fare, the unfamiliarity of reindeer (as well as caribou) allowed stories like these to fill the void of an animal that was largely unknown to most people. By filling this void with myths and literary devices, a species that was not particularly "near and dear" became over time "more of a story than an actual animal."[7]

CARIBOU IN OUR IMAGINATIONS

An oft-repeated observation about caribou is they are the bison of the North American Arctic and just as the American bison were hunted for food, clothing, shelter, and spiritual reasons, an abundance of caribou offered the people of the Arctic a way to survive. Caribou killed in the fall were dried for winter consumption, while in the spring caribou bones were crushed and boiled for the bone fat, which was then eaten along with the dried meat and fish. At Christmastime, *akutuq* or "Eskimo ice cream" was made from reindeer fat or tallow, seal oil, freshly fallen snow or water, fresh berries, and sometimes ground fish. When sled dog teams were all they had, it took a lot of caribou to feed a family and their dogs, so nothing was ever wasted. Not only was the caribou eaten, but its fur was also used for sleeping bags, mattresses, and clothing, and its hide for tent material. Caribou fur also helped waterproof the boots people relied upon to stay warm. Where the number of bison on the American plains was called "one robe," the number of barren-ground caribou in the Arctic plains was seen as "one army" or a "world [that] seemed [to be] a moving mass of caribou."[8] Ernest Thompson Seton wrote in his book called *The Arctic Plains* that the number of caribou in "[one] army was over 25,000,000" with the possible existence of other "armies." To Seton, these numbers suggested caribou "must indeed out-number the Buffalo [bison] in their palmiest epoch."[9] Since Seton wrote this in a time when caribou were already in decline, his numbers were likely an exaggeration that aligns with what some early natural historians saw as an example of his "nature faking."

By the early 1800s, the parallels between caribou and bison extended well beyond satisfying the subsistence needs of the People of the Plains and the People of the Arctic, but to an all too familiar tragedy associated with both Western expansion across the Plains and Arctic explorers searching for the Northwest Passage and eventually gold. For white explorers and settlers heading west, the need to find food, to develop a marketable resource in the form of bison coats and meat, and eventually to provide a recreated frontier experience of a bison hunt for train passengers led to the near extinction of the American bison and by association Native people roaming freely across the plains. For Arctic explorers searching for the Northwest Passage, they discovered not a passage but an ocean full of whales. In their search for oil, the presence of whales brought not only commercial whalers from all over the world but also guns, diseases, alcohol, and the whalers' sperm as well.[10] This highly diverse group of people arrived in the Arctic to make what Seth Kantner described in *A Thousand Trails Home* as the first widespread contact with Native Alaskans who, at the time, were still living off caribou as their ancestors had for hundreds of generations. As sperm whales became overhunted, the whalers looked for new ways to make money and so they moved to killing walruses for their oil and ivory, and other whale species that included bowhead whales, a whale Native people relied upon to not only make their small skin boats but to provide food and oil for light and heat.

To sustain these new arrivals, a deeply rooted traditional way of life was altered beyond recognition. Barter and then employment opportunities in a new and unfamiliar economy led the local inhabitants away from thousands of years of deeply embedded traditions. The loss of blubber and meat from whales, the disappearance of musk ox and walrus, and the tease of new trade goods encouraged Arctic people to seek employment on ships and to shift their gender and societal norms regarding men hunting caribou and women making clothes to those more useful to a market economy.

These behaviors that had revolved around family needs and were maintained by tribal boundaries limited caribou hunting areas to small family groups of traditional Arctic people. Once whalers realized caribou meat tasted better than marine mammals and that clothing made from caribou skins was much warmer than wool and oilskins, increased caribou hunting by both Native people and whalers changed traditional rules regarding how caribou could be killed. Items foreign to their way of life, including tobacco, alcohol, sugar, flour, cloth, needles, and metal, entered the trade network, disrupting the balance of traditional lifeways as well as long-established community connections. The spread of unfamiliar deadly diseases further fragmented relationships as Native communities became more and more isolated. The intense desire for fur and meat by the whalers led, in the 1870s to muskets being included as a trade item, so within a short amount of time, most Native hunters had a firearm, and thousand-year traditions were abandoned. To Kantner, "Overnight, the Stone Age became the Rifle Age."[11]

In this swiftly changing environment, the ideology of Native people in Alaska that held when a resource was "valuable and available" it was wisest to "harvest as much as possible while possible"[12] worked to their disadvantage, leading to a tragedy of the caribou commons. The persistence of killing calves for children's clothing and thermals for adults during the coldest weather compounded the assault on caribou. Yet, as with the bison, by the 1890s, when caribou were becoming harder and harder to find, caribou calves were still being harvested for their skins at calving grounds while, in 1897, Charles Brower, a famous adventurer and trader, organized hunting parties that killed twelve hundred of an already rapidly declining species.

After years of disruption from new technology, new people, and new values to Native people and their culture, some saw the arrival of missionaries in the late 1800s as a hopeful way to fill the cultural void that was left. Along with Christian values, these missionaries brought medicine, built infrastructure, and helped feed and clothe Native people whose resources had been over-exploited by hunters and traders. According to Greg Finstad, a University of Alaska-Fairbanks professor and Reindeer Research Program Manager, Sheldon Jackson, an Episcopalian minister, recognized the challenge of turning hunter-gatherers into Christians. To him, reindeer herding would be the platform for conversion to Christianity, and to gain support from the United States Congress he argued that Native people should acquire reindeer to raise their own food. Upon receiving government permission, he imported reindeer (*Rangifer tarandus*) from Russia, which, unknown to people of the nineteenth century, was the same species as caribou. Like the Plains people of the American West who, after the loss of the buffalo and their land, were put on reservations and forced to rely on support from the federal government, the Indigenous people of the Alaskan Arctic became reindeer herders or "cowboys of the north"[13] after caribou became less available. The reindeer provided people with what they needed, but the animals also brought another pathway to Western assimilation: jobs. Unlike hunting and gathering, reindeer corrals and roundups were part of an agricultural enterprise that included slaughter factories. Harvesting caribou from the wild produced free meat, while buying reindeer meat from private profit-oriented herders took cash. The turning of highly mobile Native caribou and whale hunters into sedentary Western-style business owners led Finstad to describe this change as a cultural shift from people living as "communal socialists to individual capitalists." The sharing of caribou meat across kinship groups declined when the available meat was now a commodity.

At their peak in 1932, semi-domestic reindeer in Alaska reached over 600,000. Yet in the years that followed, an increase in nonhuman predation decimated reindeer herds. Their sedentary nature, along with a calving season that occurred one-month earlier than the caribou's, made semi-domesticated reindeer calves' ideal food for hungry brown or grizzly bears leaving their dens in April and May and for wolves that appreciated a relatively stable food source. At the same time, due to less human pressure, caribou numbers were rebounding somewhat and by the 1930s were moving back into their earlier range. According to Finstad, caribou recovery, along with the migratory nature of caribou, eventually absorbed some of the more sedentary private reindeer herds into the massive caribou migrations. Through an increase in predation and the wild caribou's unintentional absorption of these privately owned reindeer into their long-distance migrations, by the 1950s, the Alaskan domestic reindeer industry was at about 25,000 reindeer.[14]

Culturally and physically, semi-domesticated reindeer and wild caribou have evolved into different animals. They have adapted to different living circumstances and are incorporated into the human experience in vastly different ways. Anthropogenic forces made semi-domestic reindeer out of artificially selected traits, while caribou relied on the forces of the environment and natural selection to adapt to their world. The processes of natural selection that created caribou genes made living in a state of semi-domestication impossible for them, just as anthropogenic forces made reindeer genes more challenging for survival in the caribou world. Unlike reindeer, or what some people call "hoofed rabbits"[15] due to their high conception and calf survival rate, caribou were susceptible to extirpation because of lower conception and recruitment rates, with far fewer calves surviving their first year. These differences resulted in vastly different outcomes for caribou and semi-domesticated reindeer. Semi-domesticated reindeer, as private property, were managed as cattle on a ranch would be in the American West, while caribou without institutions or cultural practices to protect them as a resource were treated as common property.

This common property arrangement was deeply rooted in Inupiaq[16] lore whereby the "*tuttut*" (caribou) would continue to return for the people to hunt because of an ancient deal their ancestors made with the *tuttut*.[17]

> Long, long ago, a good hunter had bad luck with the tuttut. It was because his wife had become careless with the meat. She tossed the heads onto the ground as if they were trash and allowed dirt to get into the lifeless eyes. The tuttut became frightened of the hunter. Even though he went out every day to look for food, he was not successful.
>
> Then one day when he was out on the tundra a stranger came up to him. The stranger looked at the man, but his heart was outside his chest, swinging as if on a string. It turned out this man was the chief tuttu in human form. He explained that his separate heart meant that he could not be killed even if the rest of his body died—but he and his fellow tuttut could still choose to stay away from people who did not respect them.
>
> "Your wife has to change," the chief tuttu told the man. "No more complaining, no throwing the meat or heads around." The man rushed back to his sod house. His wife listened and changed her disrespectful ways.
>
> This was the bargain between people and tuttut that we still follow today to be sure the tuttut keep coming back to be hunted.[18]

By showing respect to the caribou, the animal would always be available for the Inupiaq to hunt. Yet with the advent of guns and the motivation of profit, the story that maintained balance between the caribou and the Arctic people was lost. In Alaska, the decline in caribou numbers gave wildlife managers and government entities the opportunity to manage the caribou resource in ways unfamiliar to Indigenous people. What had once been a sustainable use of a common caribou resource turned into a tragedy when Indigenous people shifted to a market economy, where the traditional view of killing caribou for subsistence changed to a more Western ideology that killed caribou for profit. The failure of both Native and non-Native people to better manage the caribou resource provided an opening for profit seekers to substitute the cultural species that was the caribou with the commodity species that was the reindeer. An animal that for over one thousand years Arctic people could hunt for food, clothing, and rely on to maintain their cultural values of reciprocity, social cohesion, and spirituality, as well as to bank traditional knowledge for subsequent generations was being lost to them simply because an important cultural animal was being replaced with an unfamiliar marketable one. Similar failures to protect valuable cultural and ecological resources have occurred with bison, beaver, wolves, and grizzly bears.

WHAT DOES IT MEAN TO BE PROPERTY?

The profit motive from what remained of privately owned reindeer herds led to the development of an Asian market for exporting Alaskan reindeer antler velvet. To acquire the velvet, the antlers were severed from the reindeer in the summer before the antlers ossify for breeding season. Reindeer velvet, today, is still harvested by severing the antlers, which to Greg Finsted is akin to "loping off [someone's] arms" and is typically sold to Asian markets for enhancing male virility and relieving symptoms of arthritis. By the late 1970s, as the price for velvet increased, velvet became a significant part of a reindeer herder's income. In 1977, although the value of reindeer meat in Alaska was $201,380 and antler velvet $171,673,[19] the meat market remained the reindeer herder's "tundra gold"[20] and could be what one reindeer specialist called the meat flagship in any restaurant. Positive reindeer messaging relied on the perception that the meat and other reindeer products came from a pristine environment, rather than one that relied on slaughterhouses and the use of long-lasting chemicals. Reindeer herding also relied on grazing leases and permits making the connection between caribou and reindeer in Alaska and bison and cattle in the American plains a legitimate parallel.

To further the comparison between reindeer and caribou is to understand that caribou are wild animals and as public property cannot be bought or sold, while privately owned semi-domesticated reindeer as livestock can be. Private versus public status becomes more challenging when you add the presence or absence of a privately owned reindeer's earmark and its location on or off a reindeer herder's range. According to contemporary wildlife regulations, the absence of an ear tag when a reindeer leaves a herder's range to merge with a migrating caribou herd means the semi-domesticated reindeer is now understood to be a wild caribou. This distinction is important because, as private property, hunters cannot shoot reindeer, but through its association with wild caribou in a caribou herd along with the absence of an owner's identifying ear tag, the reindeer becomes part of the public trust, and as such, the hunter's property when shot. Just as the absence of an ear tag makes bison a "wild" animal that in some places can be hunted and killed, its presence defines bison as private livestock too. In the United States government's brucellosis quarantine pens just north of YNP in Montana, those bison that leave the park looking for food in the winter and are not killed in Beattie Gulch or as slaughter are shifted to pens, ear tagged, and quarantined for 660 to 1356 days. They can then be transferred through a "rehoming" process to Tribal lands in Montana. Like reindeer, the presence of an ear tag shifts a bison away from its status as a public trust resource to one of private property.

Work done by Ande Somby, a Sámi and a Faculty of Law associate professor at the University of Tromsø, sees the conflict over traditional knowledge and a reliance on cultural species to be about what happens when a "prey culture" meets a "predator culture." Predator cultures tend to be wealthy Western nations that are motivated by making money in a global economy, while prey cultures are more typically small Indigenous communities spread out over large geographic areas that are rich in natural resources.

To achieve their goals, the predator culture relies on metaphors that make it sound right for the prey culture to share or give up their land's bounty and wrong not to do so. For the Sámi, the Norwegian state relies on the "family metaphor" to explain how *all* Norwegians are in this together and each must do their part to contribute to the happy family. To Somby, this metaphor encourages Norwegians to not "stand in the way of what's best for our cozy household."[21] Another method used to maintain the predator-prey relationship is to demean the traditional lifeways and glorify a more transactional approach to land use. These approaches might include mining or shifting to big trawlers instead of continuing to rely on subsistence-style small boats. The last method is to make the people invisible. Integration into the dominant culture strips Indigenous people of their identity and alienates them from their cultural roots. If these methods aren't effective, though, the non-Native Norwegian experts will claim the problem for the Sámi is "too many reindeer."[22] Like the excessive killing of bison on the Great Plains by Western settlers and hunters was to Native Americans, this last strategy is used to control the reindeer resource to control the Sámi people. In Alaska and across Canada and the northern United States, many Indigenous people suffered the same fate when they lost access to the caribou.

It wasn't until the 1930s and 1940s, with the start of modern wildlife management, that hunting regulations came into being. These early regulations relied on seeing the caribou exclusively through the lens of biology and science, making the important cultural functions the caribou served for the Arctic people invisible. This invisibility not only alienated the Native people from their land and culture but stripped them of an essential part of their identity. To them, their identity was tied to their traditional food, and without that access, they lost an important aspect of themselves.

Although it took a new kind of "outsider" and government regulations to create a system that managed people's access to caribou, the barren-ground caribou's recovery today has helped traditional Arctic people undergo a cultural and ecological renaissance. By being swept into a kill-for-profit system, the deeply rooted tradition of land boundaries and family hunting groups offered caribou no protection from a predatory culture that, in the absence of any form of ownership, treated caribou as an open-access, commonly held resource. When these predatory outsiders arrived with guns, that technology, once it was in the hands of both Native and non-Native people, led to a significant decline in caribou numbers. The pressure on caribou from guns was compounded by the Native people's own attitude of killing caribou in large numbers whenever they were around because their origin story told them there would always be more. Small populations of Native people with less powerful weapons made that management strategy a viable one, but once the caribou story shifted away from a "property mimicking"[23] or a land boundary system that regulated and restricted access to caribou exploitation, an increase in hunting pressure from more hunters with powerful guns, along with a loss in habitat and the natural ebb and flow of the caribou's population cycle, made the animal's decline inevitable.

In the United States today, barren-ground caribou are managed by the Alaska Department of Fish and Game, with input from many constituent groups. In different parts of Alaska, as many as five caribou can be killed per day, with more stringent restrictions on whether cows and calves can be hunted at all. The official 2022–2023 hunting guide for Alaska recognized both a general hunt and a permit hunt for caribou that applied to both resident and non-resident hunters. For the permit hunt, there is a special hunt called a community subsistence harvest hunt that exists to accommodate traditional subsistence hunting practices and to create group bag limits rather than individual bag limits. Within the community of caribou managers are groups like the Western Arctic Caribou Herd Working Group. This group created a Cooperative Management Plan (CMP) that, in 2015 when caribou numbers were declining again, recognized that while restricting anyone's access to essential caribou meat was unappealing, it was still necessary for everyone "to share the burden of reducing harvests regardless of where they live[d] or why they hunt[ed] caribou."[24] This decision sounds a lot like the Norwegian government's claim that the Sámi are part of a "happy family," but as a cooperative with input from different constituencies, it has become possible for cooperation between Native people and outsiders to blend science with traditional knowledge so generations to come can benefit from sharing the land with barren-ground caribou.

THE GRAY GHOST

To understand the tremendous variety that exists for North American caribou, it is critical to recognize that caribou not only have four existing subspecies but that within those, there are many different ecotypes. Although these ecotypes share an evolutionary path that identifies them as *Rangifer tarandus*, they have developed different demographic and behavioral adaptations in response to the ecological conditions of their habitat. According to work done by F. F. Mallory and T. L. Hillis for the Seventh North American Caribou Conference in 1996, woodland caribou subspecies typically live in small, isolated herds in winter and, during the rest of the year, are relatively sedentary, migrating only short distances. They calve in the spring in bogs or islands away from the rest of the herd, frequently remaining alone until mid-winter. While this might be considered typical for woodland caribou, the George River woodland caribou herd in Quebec, Canada, with the same morphology and genetics as other woodland caribou, belongs to what was at one time the largest caribou herd in the world. Thirty years ago, that herd numbered around 800,000, but today, because of resource development (including mining and road building), contaminants, predation, hunting, and climate change, there are fewer than 7,200. Even with that decline, they still migrate hundreds of miles from boreal forest to open tundra, where they synchronously give birth over a three-week period. With the George River herd, the behavior of large migrations typically associated with most Northwest Territory barren-ground caribou is expressed with this herd of woodland caribou.[25] This variation in demographic and behavioral adaptations is based on the benefits and constraints that exist for caribou populations within specific landscapes.

Human, wolf, and bear predation, along with fire and human disturbance in the form of logging, roads, and urban development, as well as the conversion of old-growth forests to open areas ideal for moose and white-tailed deer, have led to serious declines in woodland caribou numbers. For these caribou, logging roads have reduced the advantage of calving in boreal forests because roads give wolves easier access to prime woodland caribou habitat, while the loss of old-growth boreal forests has not only attracted more moose and white-tailed deer but has also removed a low-food ecosystem barrier that kept hungry predators away. For the wolves, moose, white-tailed deer, and, at times, elk are the intended prey, but by sharing a habitat with them, caribou and their calves can also succumb to wolf predation. For woodland caribou in the southern part of Canada and the Canadian Maritimes, along with, at one time, Maine, Vermont, New York, New Hampshire, Minnesota, Washington, Montana, and Idaho in the United States, a combination of forces that included human recreation, logging, mining, and roads, along with the presence of wolves, grizzly bears, mountain lions, and possibly wolverines, contributed to what Lydia Allen, an individual actively involved in caribou recovery, called "[caribou] death by a thousand cuts." Logging roads opened access to high elevations for predators, and the construction of these roads and the presence of mining and logging camps killed caribou as a source of food. Adding to that, the fear of job loss in the timber industry if caribou habitat was off-limits for logging, pressure from the alpine recreation sector that includes alpine skiing, heliskiing, snowmobiling, and even cross-country skiing, a limited knowledge about caribou, the caribou's lack of fear compared to a white-tailed deer or moose, and a low recruitment (reproduction into the herd) rate have allowed caribou to slowly slip away. To Gregg Servheen, retired wildlife biologist and natural resource manager for the Idaho Department of Fish and Game, the scarcity of woodland caribou made it a species unfamiliar to most people, while the low numbers across a large habitat made them into "ghosts, just doing their thing."

The problem with living as ghosts was that caribou were slowly blinking out before many scientists realized they were nearly gone. Their ghost-like nature, compounded by individuals opposed to caribou in boreal forests who effectively amplified the caribou's negative impacts on their lifestyles and livelihoods, distracted scientists from appreciating how rapid the decline in woodland caribou numbers was. Many loggers are not interested in protecting caribou, while the tax revenue from logging has been used in some places to fund schools. Alpine and boreal tourism providers have concerns about land restrictions that impact their bottom lines, while high-elevation deep-country lodges rely on snowmobiles to supply these remote lodges. For caribou, the presence of repetitive direct disturbance from human activities that include high-marking snowmobilers or alpine skiers elevates the caribou's stress hormones, so they are forced to move to areas that cost them energy security and food.[26]

The absence of caribou in most people's imaginations makes caribou coexistence challenging. David Argument, a Resource Conservation Manager for Parks Canada, shared a story with me about a teacher in Canada who told her students the Canadian quarter has a moose on it, when it is a caribou. This error made Argument wonder if Canadians even know what a caribou is. This same concern was voiced by Cory DeStein, a caribou photographer, who said visitors will see eight-hundred elk, but claim they saw eight-hundred caribou. Recently, confusion over what caribou look like or where they live occurred in Yellowstone National Park (YNP) when a visitor pointed out to my husband and me what he thought was a caribou on the hillside, when it was an elk. The lack of awareness about caribou along with the inability of most people to correctly identify them challenges caribou restoration efforts, while their low profile on the landscape captured by the gray ghost metaphor makes recovery experts wonder if balancing the multiple demands on caribou habitat will have to occur behind a "really big fence."[27] For the caribou, its invisibility masks its value, while the idea of a fence as a conservation strategy suggests caribou cannot survive on a twenty-first-century landscape without human-imposed boundaries. To rely on fences, as we do for bison outside of YNP, would force us to rethink how we define wild and wildness. If the caribou is the indicator species for boreal forest and tundra health, what happens when the species can no longer engage with the land as its evolutionary relationship requires?

A reliance on the gray ghost metaphor suggests the caribou's invisibility might be part of its cultural appeal. Even after the species was designated as endangered and then placed on the endangered species list, that invisibility gave the animal space in our imaginations but made no room for it in our reality. The language of woodland caribou as ghosts cloaked it behind a boundary between what is real and what is not, so when Gregg Servheen said they were the "coolest animal," he also admitted they disappeared from the state of Idaho without the Idaho Department of Fish and Game even acknowledging the event. According to Mike Schrage, the wildlife biologist for the Fond du Lac (Ojibway) Reservation Management Division, the fungibility of caribou with huntable species like elk and moose also made caribou, which according to Schrage were never that central to the Ojibway people of Minnesota anyway, a relatively easy tradeoff. The situation for caribou today is so dire that many scientists argue that getting people behind the woodland caribou in the United States is too late because "that ship has sailed."

EFFORTS AT WOODLAND CARIBOU RECOVERY

In the United States, efforts in Minnesota and Maine, as well as Alberta and British Columbia in Canada to recover caribou have met with varying degrees of success. When Lydia Allen called the caribou loss a "death by a thousand cuts," she was referring to the nibbling away at an animal that existed in small numbers spread over large tracts of land. At the end of the 1850s, a remnant Pleistocene population of woodland caribou was scattered as far south as the middle of Idaho and even into northern New York. But as human numbers increased in Idaho, they brought farming, market hunting, and, by the 1880s, mining. The area had rich silver deposits that brought miners who stripped the forests and killed "everything and anything" to feed the men. "Man camps" for silver miners were set up in caribou habitat, so miners killed caribou as one source of food, forest fires shifted a lot of caribou habitat into moose and white-tailed deer habitat, while the presence of highly fecund deer led to increased predation on caribou as bycatch from increasing numbers of wolves, grizzly bears, and mountain lions.

Unlike the bison's high-profile disappearance from the Great Plains and YNP, the decline in caribou wasn't apparent until later in the 1900s. By the 1950s, there were laws against hunting caribou, but it wasn't until the population had dropped to approximately thirty animals in the southern Selkirk Mountain herd of extreme Northeastern Washington, northern Idaho, and southern British Columbia that an emergency ruling on January 14, 1983, first listed the woodland caribou as endangered in the United States. The recertification of that listing on October 25, 1983, identified the reasons for their decline to include poaching, habitat loss to timber harvesting and wildfires, collisions with motor vehicles, and genetic issues through inbreeding. According to the listing on the Federal Register, this was a subspecies that once occupied nearly the entire forested region from southeastern Alaska and British Columbia to Newfoundland and Nova Scotia down to the lower forty-eight states of Washington, Idaho, North Dakota, Montana, Minnesota, Wisconsin, Michigan, Vermont, New Hampshire, and Maine. Overhunting and human-induced changes to caribou habitat led to the woodland caribou's extirpation in New England by 1908 and the Great Lakes region by 1940. In the Intermountain and Mountain West, the lack of hunting rules led to a major reduction in caribou numbers by 1900.[28]

Since it wasn't until the 1930s that hunting laws were firmly established, the loss of species held in common is a story often repeated when it comes to wildlife in the United States. For the caribou, it took another thirty years before a single university professor assigned one graduate student the task of studying woodland caribou, and it was ten years later, in the 1970s, when some recommendations for improving caribou habitat were defined. But the recommendations that emerged from what people knew about caribou were based on conditions that no longer existed. Since there was no academic literature in the 1980s about predation, its role was not considered as a factor in caribou decline. Even the 1983 Federal Register Caribou Endangered Species description listing relied on the role of timber cutting only through the lens of the elimination of escape cover, migration corridors, and lichen production. Although the report highlighted the impact of human killing on the herd from legal hunting (before 1957) and ongoing poaching in areas with a lot of roads, there was no mention of the predator-prey interplay between moose, white-tailed deer, caribou, and wolves.

In the 1980s, when the caribou was first listed, the forest service was changing how it managed forests. Since these changes, which included new rules regarding old-growth forests as well as cedar and hemlock, coincided with the caribou listing, to loggers, the limits on what they could now harvest were understood to be the "caribou's fault." Along with the caribou listing was one for the grizzly bear, so it became natural for loggers to blame those two species for their new lived realities. The big green forest service vehicle became a target, and according to one forester, whenever people saw her truck, they would yell at her for protecting those "God damn caribou." In the 1980s, the USFWS didn't work collaboratively like they do today. There was no outreach to Tribes or stakeholders in the community, and no one thought to create environmental working groups, so people felt they had no voice. To this forester, the result was "an angry public" that believed, by protecting caribou habitat, the "damn caribou were taking their jobs away."

Even after the federal listing, the idea for a "save the caribou" campaign never resonated with enough people. According to Allen, grizzly bears got all the interest and for that reason, "all the money in the universe went to the Greater Yellowstone Ecosystem (GYE). They g[o]t all the love because of the [grizzly bear's] tie to Yellowstone National Park." Allen believed that focus on Yellowstone offered "squat" to the Selkirks because there were only 85–100 grizzly bears, and the animal of concern was the caribou that "did not feel the love." This lack of what Allen called "the love" is reflected in a quick Google search that indicated in 2022 there were nearly 2,500 scholarly articles written on grizzly bears, while, according to an article on current knowledge of woodland caribou in Canada produced by the National Council for Air and Stream Improvement, Inc., there was a *total* of only 320 peer-reviewed articles published on woodland caribou between the years 2009 and 2019.[29] Their absence in academic literature highlights the power of the ghost metaphor to maintain a species' invisibility (see figure 4.2).

Although the Southern Selkirks were the last holdout for woodland caribou in the lower forty-eight, there were earlier efforts in Maine and Minnesota to return caribou to the southern parts of their range. To Mark McCollough, the USFWS' Endangered Species biologist in Maine, the first major effort to restore caribou occurred in 1963 under the auspices of conservation and to "right a wrong." Although the town of Caribou in northern Maine was named in 1877 for the abundance of woodland caribou, by 1908 they were gone. To many people living in Maine, the caribou's extirpation was the result of human changes to their habitat and overharvesting from subsistence, market hunting, and unregulated sport hunting along with disease (particularly the meningeal worm parasite carried by white-tailed deer) and predation. After their extirpation, interest in returning caribou was considered many times. Initially, in 1923, Maine's governor Percival Baxter suggested reintroducing caribou to Mt. Katahdin (now in Baxter State Park); then in 1937–1938 and 1952, the USFWS proposed translocating caribou from Canada to Maine, and the Maine Cooperative Wildlife Research Unit suggested the same in 1940–1941. In 1963, the Maine Department of Inland Fisheries and Game captured twenty-three woodland caribou in Newfoundland and released them without radio tags on historic wintering habitat on the summit of Mt. Katahdin.[30] The reintroduction was a failure because, in the spring of 1964, when the animals dispersed off the mountain, they never regrouped. In 1964–1965 there were seventeen verified sightings about twenty miles from the release site, and then in 1965–1966 there were six sightings about twenty-seven miles away. After 1966, some combination of poaching, disease, and predation ended that translocation-restoration effort (see figure 4.3).[31]

The realization that woodland caribou had been native to Maine, Vermont, New Hampshire, and even parts of the Adirondacks in New York and that they were important culturally as well as a food source made the next effort at reintroduction a possibility. In January 1986, a physician who also happened

Figure 4.2 This Mountain Caribou Mural in Sandpoint, Idaho, Is Part of the Center for Biological Diversity's Endangered Species Mural Project. The goal of this project is to bring endangered wildlife onto the streets to help celebrate local endangered species within communities. Mural artists Mazatl and Joy Mallari (from the Justseeds Artists Cooperative) worked with Roger Peet to create this caribou mural. *Source*: Roger Peet. Noncommercial Sharealike 2.0 Generic. CC BY-NC-SA 2.0.

to like caribou started testing the idea of trying again to return woodland caribou to Maine. He began by gauging interest at sportsmen shows and then meeting with the Commissioner of Inland Fisheries and Wildlife. The Commissioner was enthused with the idea and quickly organized a meeting with private sector representatives, legislators, and government conservation officials to evaluate the potential for another caribou recovery effort. Within a short amount of time, Newfoundland caribou biologists were in Maine to assess potential caribou habitat. By late summer of that same year, the group decided to proceed immediately with a reintroduction. Funds were raised, with the past Maine governor Horace Hildreth contributing $50,000 to the project, and with that the Maine Caribou Project was underway. Caribou as an idea attracted tremendous media attention and gave the public a window seat on the project. According to McCollough, the public euphoria about the recovery effort led to a groundswell of support. Surrounding all this positive caribou news were the wildlife biologists who worried about the potential transmission of brainworm to caribou from white-tailed deer. Moose, Maine's existing iconic and charismatic species, already suffered tremendously from

Figure 4.3 A Young Woodland Caribou in Newfoundland, Canada, 2022. *Source*: Barbara K. Jones.

the effects of brainworm, so how could biologists support returning the caribou when they knew it would respond the same way.

Although there were concerns, the project moved forward with plans to receive up to thirty caribou from Newfoundland's Avalon Peninsula, where the woodland caribou lived in non-migratory herds. The idea was to keep a nursery herd of caribou in an environment free of white-tailed deer for five to seven years. Caribou born and raised in captivity would be released as yearlings or two-year-olds until there were approximately one-hundred individuals in the wild. The family of the artist Andrew Wyeth offered a remote island as the translocation site. Along with the deer-free nature of the island and the powerful name recognition associated with Andrew Wyeth, the project members thought this would be an ideal setting for caribou recovery. To the scientists, the project's goal was to answer the research question of whether caribou can successfully be returned post-release to an area occupied by white-tailed deer, while public enjoyment regarding the presence of a self-sustaining herd was of secondary importance. However, just before the caribou capture, a newly hired wildlife biologist decided the island was a logistical challenge, so on short notice, the ungulate research facility at the University of Maine at Orono was chosen as a temporary holding pen for the caribou. To McCollough, the prevalence of brainworm-carrying white-tailed deer around that site would eventually seal the caribou's fate.

The popularity of this caribou reintroduction idea made all aspects of the translocation from Newfoundland to Maine highly publicized. The darting of caribou from helicopters and the subsequent stress of the holding pens led to the loss of nine caribou. The caribou were transported by ferry from Newfoundland to Nova Scotia on a ferry boat called the Caribou (after a ship of the same name that had sunk) when the worst storm many people had ever experienced hit. At the ferry station, wind speeds were so high there could be no filming, but the decision was made that they had to cross. McCollough spoke of how the ship was shuddering, tables and chairs were being tossed around, and the gift shop inventory went everywhere. The boat was covered in ice, and the normal five-hour trip took many more hours to complete. The captain wasn't sure they would make it, so when they got close to Nova Scotia, he brought the boat close to shore, so, if the boat floundered, they could all swim to land. Three or four more caribou died on the ferry trip, while two more died in transport to Maine and three more shortly after their arrival. These deaths were documented and reported through an intense media presence. People were still enthused, but, according to McCollough, because people never thought any caribou would die, there were concerns about the high rates of mortality. The location of the pens at the University of Maine made public viewing opportunities possible, and within the first three years, 85,000 people visited during public viewing days. High public enthusiasm and extensive media coverage kept the caribou project in people's minds. The initial caribou recruitment success forced the pens to be expanded to accommodate their increasing numbers, with the expansion pushing the pens further into prime white-tailed deer habitat. At this point, the caribou project gave Mainers the opportunity to reconnect with woodland caribou in a bound setting. They were not competing with moose or white-tailed deer or limiting employment or recreation opportunities, so having caribou in pens offered people access without any inconvenience.

The next phase of the project required a release site, which was determined to be the privately owned Baxter State Park. The requirements for this site included no publicity about the release and that the release would occur when the park was closed for the winter. Caribou public relations efforts raised money to support the project, while an "Adopt-the-Caribou" program through the sale of T-shirts, prints, and general donations from Patagonia Equipment Company and the National Fish and Wildlife Foundation was successful in raising additional funds.[32] Despite these early successes, the intensity of media coverage that was becoming more negative in its tone and the ongoing loss of caribou were causing problems for the project. Through all that, though, public opinion remained high, and most Mainers wanted the project to continue. By 1989, the consequences of brainworm and unexpected predation by bears[33] were leading to additional caribou deaths, and the decision was eventually made to release all the caribou from the nursing pens and, rather than raise their own caribou, acquire new ones from Newfoundland each year. The final blow hit when, in January 1990, due to the controversial publicity surrounding the project, Baxter State Park decided it would no longer accept more caribou.

The project ended with the realization that wildlife reintroductions could not be exclusively private (relying on hunting groups and wealthy donors) or political (officeholders that come and go with elections) because those narrow stakeholder groups might overlook the rigorous review wildlife biologists could provide and, by that exclusion, alienate those in the professional wildlife community. The lack of academic research on caribou also made it difficult to know what could go wrong. Expanding the nursing pens into white-tailed deer habitat and then transferring the caribou to Baxter State Park, which was full of black bears, led to a sad but somewhat predictable outcome. A source of funding based on artificially high expectations and the unreliability of political support for such private projects did not allow for the time and consistent funding sources reintroductions like this can take. Although this project resulted in an increased enthusiasm for public conservation, it led to a decreased enthusiasm for future translocations of a species that appeared to live only in people's imaginations rather than as a real animal with essential habitat needs.

Assisted migration through translocations has been the norm for southern caribou herds since the 1960s. There were those attempts in Maine, and then efforts by the Ontario Ministry of Natural Resources and Forestry in 1982 to move woodland caribou from the Slate Islands in Lake Superior to Michipicoten Island off the eastern shore of Lake Superior in Ontario. At that time and up through the 1990s, as many as six-hundred-fifty Lake Superior woodland caribou roamed the Slate Islands. In 1982, some of that herd was translocated to Michipicoten Island, an island that, like the Slate Islands, had no predators. By the 2000s, four-hundred-fifty woodland caribou roamed Michipicoten Island. Yet in the bitterly cold winter of 2014, wolves walked across the ice from Pukaskwa National Park in Ontario so that in a few short years, the number of caribou plummeted. By 2018, the wolves had reduced the caribou herds to fifteen on Michipicoten and to just two males on the Slates. To save these remnant populations, wildlife advocates pushed the Ontario government to act. The government conducted emergency airlifts that winter to save the Michipicoten caribou and to restart modest herds on the Slates, where wolves were once again absent, and on the smaller, but farther-from-shore Caribou Island. According to Gord Eason, a retired biologist who was part of the Ontario Ministry of Natural Resources and Forestry reintroduction of woodland caribou to Michipicoten, in 2021 the total estimated number of caribou at both sites was an optimistic sixty animals.[34]

For caribou advocates, each time a southern herd disappears or is so greatly diminished it can't withstand pressure from any source, the southern herd boundary moves further north. That boundary movement north negatively impacts the caribou's evolved survival strategy of scarcity because there is no opportunity to spread across and reconnect with other caribou in a southern range now empty of caribou.

Through all this, the hope was that the Canadian government and the National Park Service (NPS) in the United States would work to protect the caribou. In Canada, prioritizing caribou would mean less logging and, in the United States, taking some of the wolves to Isle Royale National Park would mean less predation pressure on the remaining caribou. Decisions were slow to be made and life became dire for this population. In the late 1980s, during the rise and fall of the Slates and Michipicoten caribou populations, Jim Nelson, a math professor and avid hunter and conservationist from the University of Minnesota, wondered if it was possible to return caribou to Minnesota after a forty-year absence. The blame for their extirpation initially was placed on overhunting and habitat loss, but newer evidence suggests the meningeal worm from the expansion of white-tailed deer was a significant cause for their decline. Along with the worm, the white-tailed deer brought wolves. An already stressed and declining population couldn't deal with the predation effects of increasing wolf numbers, and compounded with the already established black bears made survival for caribou in Minnesota impossible. With Maine's "spectacular failure" as their guide, in 1988, an organization called the North Central Caribou Corporation was founded to explore the possibility of returning caribou to Minnesota. The Corporation's board of directors included representation from state, federal, Native American, and Canadian agencies as well as one citizens' organization. Like the reintroduction attempt in Maine, the effort was a private one funded by the Duluth Safari Club. It did, though, rely on research done collaboratively among the U.S. Forest Service, the Natural Resources Research Institute, and the University of Minnesota.[35]

According to John Pastor, a biologist working on this project, other sources of funds came from pull-tab contests held all over Northern Minnesota. The contests generated $50,000 to fund the initial research. The stakeholder support would have needed to include the Minnesota Department of Natural Resources (DNR), Superior National Forest, Native American groups, timber and tourism industry officials, and the Province of Ontario for the caribou stock. With an abundance of lakes and forests and few white-tailed deer, the area with the best caribou habitat was the Boundary Waters Caribou Region (BWCR) within the Boundary Waters Canoe Area Wilderness (BWCAW). Timber industry concerns were mitigated by the BWCAW's status as a wilderness area where no logging was allowed but included promises of access for areas outside that zone, while tourist opportunities would increase for canoeists interested in seeing caribou on the land. Scientists saw this project as a positive for science and for caribou because whether the project succeeded or not, the results would provide important scientific information for future restoration efforts, and, if the caribou failed to create a sustainable population, habitat deficiencies critical to their survival could be identified.

After two years of working to create a model for all future reintroductions, the project was shot down by the DNR. The scientists working on the project were shocked that after all their work, the DNR saw the scientists as stakeholders presenting evidence rather than scientists presenting facts. The scientists working to bring caribou to Minnesota were told to "compromise [their] science." According to Pastor, logging and white-tailed deer hunting groups determine everything in northern Minnesota. Timber sales, mining, and white-tailed deer hunting are the big sources of revenue for the DNR, so there was pressure not to return caribou because, with caribou on a shared landscape, the white-tailed deer couldn't expand their range. As one of the most powerful hunting organizations, the Deer Hunters Association went to the capital and demanded protection for white-tailed deer. Clear cutting for pulp generated aspen for loggers and great habitat for deer, but terrible habitat for caribou. To Pastor, caribou are so far away from most people that they are not on the public's mind, and this absence of awareness has generated few advocates for caribou in Minnesota. Curiously, by bringing more people out into the woods, Pastor believes Covid-19 has changed that. He thinks people are paying attention more now, and even school kids are being taught to protect what they love, but without a fan base driven to protect caribou, their status as a ghost in the forest makes that love harder to develop.

WHAT'S NEXT FOR WOODLAND CARIBOU?

The caribou story in the forty-eight contiguous United States ended when the last few members of the Southern Selkirk herd were relocated back to Canada in 2019. According to Gregg Servheen, woodland caribou are a large charismatic game animal, yet their story was not like an American movie where there is a happy ending. Even after their endangered species listing, there were only a few people who worried about the woodland caribou, so the animal disappeared without a whimper. Servheen believed that since the decline in caribou numbers coincided with a decline in sturgeon, grizzly bears, and lynx, to those living in the Bonners Ferry region of Idaho, the caribou listing was just another "straw on the camel's back." Their historically low numbers, as well as a dispersal strategy that protected them from predation, made the caribou invisible to most people. The timber and recreation industries, specifically snowmobilers, were opposed to restoring caribou because they saw their presence as a blow to their livelihoods, while Servheen's weekly meetings with local people made it clear that those living and working in caribou habitat were not interested in caribou recovery. To him, if the process of generating caribou awareness had happened earlier and was more inclusive, the public's response to caribou might have been much different.

The loss of woodland caribou in the United States is an example of human hubris because by waiting so long to recognize their needs, the trends today might be what one wildlife biologist claimed are "too big to buck." Climate change and the money needed to maintain caribou habitat in an area that has grown to rely on the economics of logging, white-tailed deer hunting, and alpine recreation have made caribou recovery particularly challenging. For animals like the caribou that tend to live in very politically conservative areas, the current "Balkanization" of politics has made the idea of wildlife recovery understood through a highly partisan political lens. To Servheen, tourists might drive through Glacier National Park and see there is plenty of nature, so when they drive home to Levittown, Pennsylvania, they might not think any more about the plight of "all this nature." For caribou that are understood to be ghosts, this absence of thought in most people's minds makes sense. It is an animal that has not encouraged a fan base, so its voice is absent in academic research, ecotourism opportunities, and local enthusiasm.

A trip to Nakusp and Revelstoke in British Columbia to study caribou recovery is a reminder of the many missed opportunities that have existed to help shift the caribou story away from one of myth to one of lived reality. Just outside the town of Nakusp, there are caribou maternity pens where caribou enthusiasts are actively working to recover caribou. Yet, in town, there is nothing that reminds a visitor or even a local person that something amazing is happening in this community. There are no caribou stuffed animals or T-shirts that highlight the work of these caribou volunteers, so even in Nakusp, generating enthusiasm for this unique large charismatic species that lives in the forests around their town can be difficult. To build a sustainable future for woodland caribou and the people who share its habitat, communities like Nakusp and Revelstoke need to consider developing ecotourism opportunities that can provide not only new revenue streams to offset losses from any limits placed on logging and alpine recreation but also a way to educate residents and visitors alike about how, by protecting caribou, they are protecting boreal forests that, as carbon sinks, can mitigate climate change. To help in that process, the woodland caribou can't be just a ghost in the forest, but rather an animal that is valued, and its restoration makes sense to caribou fans from Nakusp to Levittown (see figure 4.4).

According to Cory DeStein, a director for the Arrow Lakes Caribou Society in British Columbia, Canada, and a professional photographer, woodland

Figure 4.4 Arrow Lakes Caribou Society Maternity Pens, July 2022. *Source*: Morgan Kelly.

caribou live in small pockets, so they suffer from a public relations problem. This problem fails to consider their many amazing adaptations for survival in a boreal ecosystem. They live off arboreal lichen, which he says is like "living off a cup of noodles when you are running a marathon," and they recycle their urine to maximize protein capture. Even though they are highly adapted survivalists, they currently are better at dying. Because they are not "flight" animals, they act almost like domesticated animals on roads in the Kootenai Pass, so with the increase in road traffic and number of wolves, their skillset is not well-suited for such a changing landscape.

In addition to the forces of climate change, social and economic conflicts also make caribou survival challenging. The rise in heliskiing has put tremendous pressure on caribou because, with their poor diets, disturbances from skiers can cause them to run and waste critical calories or fall off cliffs. They can also get chased into avalanche country where one avalanche can wipe out an entire herd. Due to climate change, fire is occurring more frequently, while predation is also an issue, with mountain lions and gray wolves being significant threats. In the United States, where wolves were successfully reintroduced to the Intermountain West in 1995, there is a tendency for many Americans to romanticize and anthropomorphize wolves so that a New York City resident will write a $1000 check to save wolves when, according to DeStein, wolves will recover, but caribou will not. He thinks people are uncomfortable seeing one animal die for another, yet killing predators might just be a quicker short-term solution than repairing habitat or changing our behaviors, which takes significantly more effort and time that woodland caribou donot have.

According to a wildlife biologist for the Kalispel Tribe in Washington, there were so many management levers that had to be lifted to get anything done for caribou. Since they crossed international boundaries and were impacted by so many factors, caribou recovery required a holistic approach rather than the more typical piecemeal government agency management style. The emergency nature of the woodland caribou's needs also required the "heavy lift" (expense and difficulty) of getting understaffed and underfunded agencies to work together against an absence of political will. Although not necessarily shared by all caribou biologists, this biologist believes the true tipping point for caribou was not the factors that initially led to the caribou's decline, but the inability of agencies to better manage predators. Before wolves, it was cougars (commonly called mountain lions) that kept caribou recruitment at such low numbers. One cougar in Washington called "Mr. Nasty" was potentially responsible for hastening the end of caribou in the state and yet, after killing five cows and calves, no state wildlife agency would remove the cougar from caribou habitat. I was told the easiest and quickest way to hunt cougars is with hounds, yet a 1996 Washington law made hunting with hounds illegal, so because of the poor optics associated with breaking their own state law, the agency that could have removed the cougar, didn't. Eventually, though, the cougar was killed in Canada by a legal hunter, but by then it was too late for Washington's caribou. Even requests to collar wolves so that those managing caribou knew where the wolves were going went unheeded. The high bar required for caribou recovery let the failure to coalesce around a shared caribou effort be the answer. For ten years, the Kalispel Tribe sounded the alarm, but all they got back was the sound of "crickets," and with this failure to listen, caribou in the United States, that after 10,000 years of coexisting with other species, disappeared in eighty.

To address these social and economic struggles, affected stakeholders in Nakusp, including First Nations, loggers, and recreationists are working together to come up with solutions to this caribou challenge. For their efforts, they devised maternity pens where cows could give birth in safety and their calves could grow for their first few months without fear of predation. In the first year, 2021–2022, they captured seven cows and a yearling. The pregnant caribou cows are typically captured in March and the calves are born in June. To ensure healthy caribou upon release, they are fed a combination of nutrient-rich pellets and lichen collected by volunteers and school children so that by late July, after the calves have had some time to mature, both the cows and calves are liberated from the pen in time for mating season. In January of 2023, five of the six calves were still alive, an amazing success rate in a world full of dangers to caribou.

In coordination with addressing the caribou's biological needs, there now is a communication plan that includes promotional videos, with one shown at the Nakusp Hot Springs and others on YouTube, the sale of wildlife photos and caribou prints, and fundraisers that include a caribou calendar and a GoFundMe page, and social media efforts like the Arrow Lakes Caribou Society Facebook page, a lichen picking page, and a maternity pen page. They have also held open houses so people could come see the caribou. As part of a larger stakeholder effort on June 6, 2022, the first World Caribou Day was held. The event was created by the Caribou Conservation Breeding Foundation to educate and engage the public in helping this species at risk. These efforts, working in conjunction with affected local communities, can only help move the caribou into more people's reality (see figure 4.5).

In Newfoundland (NFL), after a caribou road trip that took me from St. John's in the east to Gros Morne National Park in the west, I was fortunate to spot one woodland caribou just north of the national park. There were caribou hoofprints on a trail at Western Brook Pond, and a caribou crossing warning sign on the main road through the park, yet there were very few people I spoke with who had recently seen a caribou. One man saw a caribou on a

Figure 4.5 Arrow Lakes Caribou Society Lichen Harvest, 2021. *Source*: Arrow Lakes Caribou Society.

hillside near his home several years ago, another said she had seen some caribou years before on a grassy knoll not far from Western Brook Pond, while one couple remembered seeing caribou when out hunting for moose. At a gas station north of Gros Morne National Park, the cashier at the gas station suggested a visit to Port au Choix National Historic Site where caribou tended to rest on the gravel near the lighthouse. It was there that I spotted my first and only island woodland caribou. According to one hunter, although caribou are one of the few native mammals in NFL, their numbers are in decline, while non-native moose numbers have increased significantly since they were introduced to the island in 1906. He wasn't sure why caribou numbers were in decline, but he did say that caribou, like puffins, weren't easily frightened. You could walk up to a puffin and "hit it with a board because they don't fly away," while caribou won't flee when you come too close. That behavior can't enhance their survival.

In Idaho, efforts at recovering woodland caribou began in the 1970s with habitat recommendations. The International Mountain Caribou Technical Committee (IMCTC) was established in 1994 to address the 1993 caribou's Blue List "species at risk" designation in British Columbia and the Endangered species designation in the United States in 1983. Through the auspices of Canada's Federal Species at Risk Act, the Committee on the Status of Endangered Wildlife in Canada (COSEWIC) designated caribou in the Southern Mountains National Ecological Area (SMNEA), including all Mountain Caribou, on their Threatened list in May 2000 and reaffirmed that designation again in May 2002. According to a report produced by the Mountain Caribou Technical Advisory Committee called *A Strategy for the Recovery of Mountain Caribou in British Columbia,* the transboundary nature of this small, local population of woodland caribou made their recovery one of provincial, national, and international significance. Yet all these designations did not save the caribou in Idaho.

The Southern Mountain caribou comprises up to fifteen sub-populations that are unique in that they can occupy mountainous regions of the inland temperate rainforest from east-central British Columbia to the inland northwestern United States. Unlike other woodland caribou, the Southern Mountain caribou do not make long-distance migrations; instead, they practice a form of movement that involves migrating vertically up and down the mountains several times a year, seeking food and cover. The South Selkirk sub-population was the only sub-population of Southern Mountain caribou that ranged into the contiguous United States. This group is called the "deep-snow mountain" ecotype because its members remain high in the mountains in the winter where the deep snowpack allows them to forage on arboreal hair lichens. Sub-populations of Southern Mountain caribou are typically comprised of multiple herds, but in the case of the South Selkirk sub-population, only one herd represents the entire sub-population.

From a recovery standpoint, the Southern herd increased, due to transplant efforts, from twenty-five animals in the 1980s to forty-six animals in 2008–2009. According to Noah Greenwald, Endangered Species Director at the Center for Biological Diversity, in the 1990s, when the USFWS maintained a caribou augmentation program, the number of caribou increased to a high of around one-hundred animals, but then the agency "completely dropped the ball" on augmentation, and caribou numbers again declined. In addition, by allocating only 30,000 of the 300,000 acres of critical caribou habitat the USFWS promised, Greenwald noted that the agency showed it "didn't want to stick [its] neck out to get anyone [opposed to caribou recovery] upset." Forest fragmentation through the construction of highways not only created a barrier to movement, but salting roads put more salt-loving caribou in harm's way, while fires, human disturbance, and increased predation also chipped away at their numbers, so a 2018 count found only three females left. In 2019, the decline led to the last South Selkirk caribou in the United States being returned to British Columbia.

Although the South Selkirk sub-population is now gone in the United States, there is hope for the Southern Mountain caribou population in Canada. In the United States, woodland caribou recovery through a 2019 outline for a caribou reintroduction is being considered (although, according to the 2022 draft of the *Periodic Status Review for the Woodland Caribou in Washington*, without human intervention through a form of assisted migration, the outlook for a caribou herd in the state is probably poor), while in the provinces of Alberta and British Columbia efforts are underway to help these woodland caribou recover. The maternity pen in Nakusp, British Columbia, is one approach in an area that has enough caribou cows to capture and protect in secure pens. According to Cory DeStein, there are caribou "groupies" who care about individual caribou, but because of the animal's invisibility, they "assume" there are another 100,000 caribou somewhere "out there." This assumption has led alpine recreationists that include snowmobilers, cat and heliskiers, and other stakeholder groups, to worry less about how their behaviors impact the caribou's fate, so that "throw[ing] a stitch on a hemorrhage while waiting for the government to save the caribou" is seen as doing enough. The invisibility of the caribou's ghost-like nature makes working with First Nations and Native Americans so critical. To DeStein, the Kalispel Tribe in Washington state is one of the strongest supporters of woodland caribou recovery.

In Jasper National Park in the Canadian province of Alberta, another effort to recover caribou is underway. Archaeological evidence indicates that First Nations people and larger and stronger herds of caribou coexisted in the region for thousands of years. By the 1950s, there were as many as seven hundred caribou in and around the park, but since then, caribou numbers have significantly declined. According to David Argument, past management decisions put the park region in this current crisis. When the park was established in 1907, Europeans introduced wildlife management practices that were very different from those of Native people. Park wardens controlled fires and, until 1959, killed predators, so wolf numbers declined. In 1920, elk were reintroduced and with low wolf numbers, the elk population grew to be significantly higher than their historical numbers. Moreover, a change in attitude about wolves on the landscape ended the wolf-kill policy, so existing high numbers of elk, along with the practice of dumping wildlife kills in the park, led to a rapid increase in wolves with caribou succumbing to predation as part of the mix. Then, in the 1970s, alpine recreation in caribou territory became more popular, putting additional downward pressure on caribou numbers.

By 2000, park management changed to slow caribou declines. Of the four local herds on the land in and around Jasper and Banff National Parks, between 2009 and 2018, two of the four herds were lost, leaving less than fifteen caribou with only three females in one herd and less than fifty-five with ten females in the other. According to Argument, the quasi-extinction threshold is less than ten caribou. The fear of extinction led Devon Kelly, a speaker at the June 22, 2022, public comment session for woodland caribou recovery in Jasper National Park, to say no Canadian wants to see any more native species go extinct. To Kelly and others, the successful recovery of bison across Canada made working to increase caribou numbers something that now seemed possible. The caribou recovery options included doing nothing, maternity penning, translocations, or conservation breeding. Since doing nothing was not a reasonable option, and unlike in Nakusp, maternity pens and translocation were not possible because in some areas there were no caribou for capture, that left conservation breeding, which would allow caribou to reproduce in a breeding facility and then be released in late fall when grizzly bears had denned, and the wolves had moved out of the area. Even with that option, fire and chronic wasting disease (CWD) are still challenges that need to be addressed when determining sites for breeding facilities.

Successful recoveries have relied on public engagement that gives affected stakeholders a voice. That effort would include using the bison recovery in Canada as a model for increasing caribou advocacy and awareness, encouraging school children to collect lichen for breeding facilities, managing wolf numbers in caribou habitat, and creating mobile education trailers to broaden the caribou's appeal. For the Jasper caribou recovery to work, stakeholders will have to tolerate recreation area closures, wolf culls, and, because Jasper National Park is not a fenced park, a willingness to accept caribou freely moving across boundaries. A successful recovery exists when the captive caribou become self-sustaining members in blended wild and captive herds.

As ghosts, woodland caribou face tremendous obstacles. Their invisibility makes learning to live with and love them difficult, but by encouraging coexistence through education and increased awareness, the language of the gray ghost that initially defined their invisibility could instead highlight their scarcity and become a force for their recovery. Unlike bison that in most places continue to be held behind fences, the absence of fences will allow the caribou to spread out and "get into trouble," which, to Argument, is a sign of recovery success. A willingness to share space with woodland caribou brings them into our lives and imaginations, making them less ghost-like and subsequently more valued. So, when a Kalispel Tribe wildlife biologist calls the caribou "a canary in the coalmine species," we should be reminded that their presence on the land gives pikas, marmots, and many other species the necessary habitat for survival, while their successful recovery offers us a second chance to live with them on an unbounded and shared landscape.

NOTES

1. 178.
2. Gregg Servheen personal communication February 28, 2023.
3. Worlddeer.org.
4. Abbots Bromley Horn Dance.
5. Moore, C.C., 1848.
6. With Rudolph added in 1939 as a promotional Christmas coloring book character for a large U.S. department store.
7. Gregg Servheen personal communication, March 4, 2022.
8. Seton 182.
9. Seton 182.
10. Kantner 60.
11. Kantner 62.
12. Kantner 139.
13. Kantner 141.
14. Greg Finstad, personal communication April 13, 2021, Stern et al. 48.
15. Greg Finstad, April 13, 2021.
16. Members of the larger Inuit culture, which spans the entire northern coasts of Alaska and Canada, as well as Greenland.
17. caribou.
18. Partnow 7. A retelling of "The Chief of the Caribou" as told by Elijah Kakinya in Nunamiut Unipkaanich.
19. Stern et al. 68.
20. Greg Finstad, personal communication April 13, 2021.

21. This is evident with efforts by the Norwegian government to achieve carbon neutrality by 2050 through the installation of "sky-high" wind turbines on land used by Sámi reindeer herders. The disruption of their reindeer from these turbines tells the Sámi, "Norway does not value the Indigenous population very highly." Taraldsen, Paulsson, and Starn. February 21, 2022.

22. Schwartz 109–111.

23. McCay and Acheson 5.

24. Western Arctic Caribou Herd Working Group 1.

25. Mallory and Hillis 49–50.

26. Gregg Servheen personal communication February 28, 2023.

27. Suzuki, March 21, 2013.

28. Federal Register October 25, 1983.

29. NCASI. 1.

30. McCoullough 1991: 2.

31. McCoullough 1990: 2.

32. McCoullough 1991: 6.

33. Lack of any academic research on caribou predation caught those involved in the caribou project off-guard because no one knew bears could run down moose, deer, and caribou.

34. McGlashen. August 7, 2021.

35. Jordan 1996.

Chapter 5

The Gray Wolf and the Grizzly Bear

Little Red Riding Hood *and the* Teddy Bear

WHAT IS IT WITH PREDATORS?

To develop a shared understanding of wildlife's place in a human world requires establishing a relationship based on culturally accepted values. For wolves in North America, the human-wolf relationship initially was a companionable one, and since wolves and people didn't compete for resources the place for wolves was not rigidly defined. This flexibility allowed wolves and humans to move around the landscape in a relatively non-hostile way. As co-members of a shared human-wildlife community, Native people living in America pre-European settlement created stories and myths that recognized a more circular relationship with wolves, where neither they nor the wolves were necessarily dominant. From this, the Cherokees created a wolf clan, the Arapaho recognized wolves as teachers, and the Sioux saw wolves as warriors. In sharing the land, Native people coexisted with wolves.

Many Native American tribes saw wolves as a critical part of the landscape and their lives. To the Shoshone and Arapaho Tribes, wolves showed them how to care for both their young and old, how a hunting pack style was safer, how no part of an animal should be wasted, and how it was everyone's right to wander to wherever their food happened to be. According to their stories, wolves, like the storytellers themselves, did not know boundaries, yet over time wolves were constrained by boundaries just as Native Americans were on reservations.[1] Even today, in Wyoming, the Wind River Reservation's Wolf Management Plan includes traditional tribal views of wolves "as kin, as strong, as deserving of respect and placed here by the Creator for a purpose."[2]

To the Sioux, although a warrior was someone who sacrificed himself for the good of others, a wolf was also considered to be a warrior. Both the wolf and the warrior had the stamina and stoicism that were valued qualities in war. Barry Lopez wrote in *Of Wolves and Men* how Sioux warriors would tie a wolf tail around their lower leg or ankle or to the back of a moccasin to signify a battle accomplishment. For the Mandan Tribe, there were specific rules about when and how many wolf tails a warrior could display depending on his role in the outcome of the fight. There were wolf songs, wolf soldiers, and wolf men who dressed as wolves; there were even times when men, by "having the mindset of the wolf," could "*be* [a] [w]olf."[3] To the Pawnee, the wolf was an animal that moved "like liquid across the plains: silent, without effort, but with purpose."[4] For a warrior in enemy territory, it was essential to move like a wolf, and that meant "becoming the wolf." Their tracking ability was based on a deep knowledge of the environment, so the boundary between the warrior's sense of self and the wolf was not always clear.

Wolf iconography was so deeply embedded in some Native people's worldviews that they ate wolf berries, played wolf chase and a game of lots called wolf sticks, and used wolf moss for dye. To them, the human world and the natural world were integrated into one whole. This belief conflicted with the settlers' relationship with nature, where many settlers were afraid of and separated from the natural world. When Europeans first arrived in North America, there was plenty of space for both wolves and these early immigrants who were mostly fishermen, explorers, hunters, and trappers Any negative feelings they had towards wolves did not initially define their presence on this resource-rich landscape; yet, with the arrival of more Europeans and their horses, cattle, pigs, and sheep, a preexisting anti-wolf sentiment resurfaced. The fear of wolves ravaging the bodies of settlers who died on the trails West made hasty burials the norm, while the threat of wolves eating their relatively docile and unprotected livestock reinvigorated a wolf loathing long established in Europe. To one Wyomingite, white European cultural attitudes rooted in stories where the wolf was the vilest antagonist made killing these animals seem reasonable and natural. A worldview forged by long-lasting anti-wolf tales, including those found in the Bible, the childhood classics of *Little Red Riding Hood*, and *The Wolf and the Fox*, as well as those that depicted wolves as shape-shifting man-wolves (or lycanthropes) encouraged and maintained an adversarial relationship with wolves because they not only frightened the settlers, but wolves also took the byproduct of their productive labor.[5] Evidence of this hatred was further demonstrated by the way soldiers treated the Cheyenne dead who were intentionally placed on burial scaffolds as a sign of respect. Since soldiers believed this was done to protect the dead from scavenging wolves, to display their disdain for both the Cheyenne and the wolf, the scaffolds were torn down so the wolves could desecrate the bodies.[6]

The eradication of wolves that began as a campaign by private landowners and farmers to rid themselves of wolves turned into a bounty-driven, state-sponsored profession. In 1814, in retaliation for killing many of his sheep and colts, a midwestern farmer spent the winter building pits as wolf traps. To him, the wolves had taken his property, and he was going to pay the thieves back in full. As with other early conservationists whose conservation bonafides would be disturbing today, John James Audubon was in attendance and witnessed a pack of dogs rip apart the trapped wolves. Wolf killing was a sport that apparently both he and the farmer found enjoyable to watch because it fit in with the thinking of the day that not only did wolves not deserve life but they needed to suffer in death.[7] This hatred of wolves justified setting them on fire, tearing their jaws out and cutting their Achilles tendons, poisoning them with strychnine, arsenic, and cyanide, or baiting them with meat-infused fish hooks. At times, this wolf-killing fever reached such intensity that other animals became victims of their poison campaigns, with the wolf killers even occasionally poisoning themselves. Wolf haters or wolfers became so unhinged that when they torched the woods to get rid of wolf havens, their own property could be lost.[8] According to a February 25, 1915, article called "More About the Wolf Bounty" in the *Jackson Hole Courier* (Wyoming), residents were encouraged to support the formation of a booster organization to exterminate the wolves. The feeling was if all the businessmen and ranchers supported "go[ing] after the wolves," the general prosperity of Jackson Hole would improve. On into the twenty-first century, the wolfer's behavior has not improved much, with shooting wolves from airplanes, chasing them with snowmobiles, raiding their dens, baiting them out of parks, or choking them to death with snares are still allowed in several American states.

In 1877, in Yellowstone National Park (YNP), freelance wolfers poisoned ungulate carcasses with strychnine, such that by 1880, when YNP Superintendent [Philetus] Norris wrote in his annual report how the hide value of the "large ferocious gray or buffalo wolf" and "the sneaking and snarling coyote" and their easy slaughter from strychnine have led to their near extinction, he didn't sound particularly remorseful.[9] Since wolves had no place in productive agriculture, killing them became part of the process of turning what Cotton Mather called the "howling wilderness" into a "fruitful field."[10] That transition required making the wolf into the face of the wilderness, which meant to get rid of the wilderness you had to kill the wolf. This logic fit in with the military metaphors early pioneers used that suggested that to conquer the wilderness, it would require pushing back against chaos and subduing the earth.[11] In the 1630s, when Roger Williams called the wolf a "'fierce, bloodsucking persecutor' of everything that was high-born in man" or Theodore Roosevelt in the early 1900s argued that the wolf was a threat to progress because it was "the beast of waste and desolation,"[12] they were defending an agrarian ideal that identified the industry of clearing the land to raise cattle or crops as a human right.

To maintain this wolf hatred depended on institutions like the church that encouraged productive labor, the government that supported settlement, a concept of private property, and a legal system that legitimized wolf killing as well as wolfer's stories and myths that could be passed down through time. According to Jon T. Coleman in his book *Vicious: Wolves and Men in America,* the tenacious devices of private property and folklore allowed Euro-Americans to claim the land at the expense of the wolf.[13] With metaphor, the wolf could become the face of the settlers' fears and anxieties about their family's legacy as well as the unknown that came from living in an unpredictable space. The wolf moved from being a real animal made of blood and bone to a "folk villain"[14] that evoked meaning far exceeding its potential to fulfill.

By the late 1800s into the early 1900s, the overexploitation of bison, elk, deer, and moose by settlers in the American West had depleted the wolf's natural prey. That loss, combined with an increase in wolf numbers due to years of readily available bison carcasses left to rot by wasteful hunters, pushed wolves to prey on livestock instead. Ranchers, unhappy with this, resorted to professional wolfers and bounty hunters to eradicate the wolves. By 1885, the increasing demand for land and the power of wealthy livestock owners led the federal government to establish the U.S. Bureau of Biological Survey. This agency was initially tasked with the benign agenda of researching birds and insects but eventually became the center for culling wildlife, particularly wolves. To encourage wolf culling, the livestock lobby relied on socially amplifying the presence of wolves by saying their land was "infested with wolves" or it was a wolf "breeding ground."[15] A government-sponsored bounty program that lasted until 1965 offered $20–$50 to anyone who killed a wolf through any means available. By baiting wolves with poisoned animal carcasses, the unintentional losses were significant in that eagles, foxes, bears, and ravens, as well as any other animals that fed on the poisoned remains, were dying along with the wolves.

During the Civil War, General Ulysses S. Grant wrote how, judging from the noise the wolves around his unit made, there must have been a "pack of twenty wolves" nearby. When the wolves were finally spotted, it turned out the twenty were just two.[16] This behavioral tool, called the Beau Geste Effect[17] is used by wolves to make their packs sound larger than they really are. A 2017 study by Kyle Brewster and his colleagues at Texas A&M on the "Human Perception of Coyote Group Size" found that coyotes also rely on this tool to distort what listeners hear. After surveying four hundred and twenty-seven people, the researchers determined that people overestimated the number of coyotes by a factor of two.[18] Social amplification can be an innocent exaggeration or part of a larger agenda. If you want more money for pest eradication, a powerful message might be to overestimate the number of birds eating up your fields or laying waste to your crops. For wolves, a deeply rooted wolf hatred compounded with a tendency to socially amplify their presence continued into the 1920s (or 1940s, depending on where you were) until the brutality ceased to be relevant because the wolf had been extirpated across most of the lower forty-eight states. Their absence and the aggrieved individual's swing away from personal wolf vendettas to professional federal agents allowed wolf stories and legends to shift from wolves as vermin to what Coleman called "exquisite vermin," and considering "exquisite" can mean anything from beautiful to elegant, the shift was significant. This switch helped the wolf become something akin to the doomed underdog that heroically fought to the end.[19] This new wolf story, along with the inevitable return of wolves on their own from Canada, encouraged a compromise from ranchers to allow wolves, at great public expense, to be returned to YNP in 1995. To welcome the wolves' return, school children waited at the Roosevelt Arch at the northern entrance of the park, waving scarves and cheering. Unfortunately, for some ranchers and hunters in Montana, Idaho, and Wyoming, this acceptance was short-lived.

Memories are short yet tales are long, so that today's wolfers, who in most cases weren't even born or at most were young children when the last wolf in the Intermountain West was killed, believe the wolves reintroduced to YNP were not the same wolves as those their forebearers had eradicated. They call the old wolves prairie wolves when prairie wolves are coyotes. One elk hunter said the reintroduced wolves were Canadian gray wolves while Montana's wolves were smaller Minnesota timber wolves. This could be believable if wolf-kill photos from the nineteenth and twentieth centuries didn't show just how large these wolves were. Without evidence, many ranchers are also convinced that a dead calf is the work of a wolf when it is far more likely the result of a feral dog. Hunters argue that wolves are to blame for plummeting elk numbers and that wolves have chased elk out of the high country, which makes traveling into the mountains to kill an elk as their grandfather or father did no longer possible. But like the wolf stories, this elk narrative is inaccurate as well. Since elk are plains, forests, and riverbed animals, their presence across a wide range of landscapes diminishes the idea that elk are moving out of the woods and into the ranch and open land due to wolves. It is also flawed because it fails to consider the hunter's own role in elk movement. During elk hunting season, elk tend to move off public land onto private land, suggesting it is the human hunter that has contributed to the perceived shift in elk behavior. As elk move onto private land (land that is usually open only to hunters with hunting guides), local hunters feel they are left out of some of the best elk hunting spots. Wolf enmity has made it possible to believe that, like the dead calf or the lack of access to prime elk hunting grounds, the obvious villain must be the wolf.

When social scientists use the word "other" to explain how people distort someone or something unfamiliar to them, they are reminding us of that near-universal human tendency to assign negative attributes to the behaviors of people or animals they don't like or understand. Whether it is through fear or ignorance, it is easy to make the familiar more foreign through adages that maintain the fear and rationalize the ignorance. For predators, their skills at surviving are admirable and their ability to challenge us in unexpected ways makes them a worthy adversary; yet, by vilifying their successes and exaggerating or misrepresenting their reality through frightening and anthropomorphized stories and metaphors, it becomes easier to not only deny them life but to subject them to a terrible death.

"SAWDUST IS MAN GLITTER"

So much of our relationship with wildlife is tied to history and the stories that flow from that history. For nature and wildlife in the United States, the history that came to define this relationship relied on seventeenth-century European settlers who saw America's natural capital as unlimited and bountiful. This relationship was also tied to a set of accepted gender roles and a way of seeing the land's value through the lens of productivity, where taking or producing something from the land was its highest and best use. Since men were the hunters and trappers as well as the sex that typically husbanded the land, an understanding of male labor that included hunting, trapping, and working the land became closely tied to notions of masculinity. Once the land was tamed and much of the wildlife extirpated, in urban and suburban areas, particularly on the east and west coasts, the connection between masculinity and working the land or hunting and trapping weakened. However, in rural communities throughout the United States, that relationship persisted and has, in part, made forging a shared American environmental ethic difficult to achieve. This history suggests American masculinity is closely aligned with controlling nature through domination and the cultural persistence of earlier settler streams.

In Ann Stinson's memoir, *the ground at my feet: sustaining a family and a forest*, the connection between productive land use and masculinity is an integral piece of her narrative. The inability to describe what it means to be a tree farmer led her, at age nine, to keep the sight of tree stumps away from her friends because, in her mind, "city people looking at a cutover only see destruction."[20] To her, the transactional space between her rural neighbors who work the forest and the urbanites who consume the forest's product is a "secret and silent"[21] one rather than a public place to forge connections between producers and consumers. This consumptive use of nature's largesse can be a challenging space to fill in part because too many competing cultural issues are part of the space's weave. When a forester compares the cutting of trees to a hunter killing a bear in that loving the trees and the bear makes sense in their respective deaths, the analogy lays bare the divide. The divide is further enhanced by the very idea of farming as husbandry, where men look after the household and the land. In Stinson's book, the saying on a forester's coffee cup that reads "sawdust is man glitter" expresses the idea that taking from the land is something a man does.

The gender of productive land use, whether in the form of extraction or husbandry, is tied in many ways to the identity of the West. When biologists attempted to transplant grizzly bears into Montana's Bitterroot Ecosystem in the 1990s, environmental historian Michael Dax described the battle as one between the Old West and the New West, where the Old West operated with the idea that "subduing and conquering nature was necessary to lay claim to masculinity and to ward off modernity's corrupting influences," while the New West believed that restoring grizzly bears (or that wildlife had rights that might supersede ours) was "a defining masculine and authentic experience."[22] By seeing conservation through the lens of gender, finding a shared view becomes even more challenging since it is intimately attached to an individual's identity. Dr. David Mattson of *Grizzly Times* adds to this point when he argues that hunters, who he sees as overwhelmingly non-Hispanic white males, view wildlife as an "agricultural crop" that generates a "harvestable surplus."[23] To Mattson, this view was part of the early conservationists' idea that big game referred to large-bodied herbivores that formed the "stock" available to sportsmen for trophies and meat. Predators like bears, lions, and wolves competed with that "stock" and as such needed to be eliminated.[24]

Hunting as an extractive practice not only fits into a masculine versus feminine dichotomy but the rural versus urban, conservative versus liberal, and the production versus protection/conservation dichotomies. Historically, it was hunters who, through the absence of rules and ownership—an ideal setting for tragedies of the commons to occur—killed what could be killed. By 1851, beavers were practically extirpated in the lower forty-eight states, followed by bison almost disappearing by 1887. By the 1920s, the wolf existed only in Minnesota and Michigan of the lower forty-eight states, while the passenger pigeon was gone forever. Grizzly bears that in 1831 "were everywhere"[25] in California's Napa Valley were, by the late nineteenth century, almost hunted out of existence. The last elk east of the Mississippi River were killed in 1865, while white-tailed deer almost disappeared in Virginia by the 1940s. This intensity of killing was the result of hunting zealots who tended to be men. As early as 1769, an act proposed in Colonial New Jersey to preserve deer numbers suggested there existed both a concern over the decline in deer numbers and the way in which deer were taken. To mitigate these concerns, hunting seasons and regulations about how and where deer could be killed were passed. A new elite group of men interested in sport hunting pushed for hunting restrictions on "waste and unimproved land" to those with enough personal wealth (over one-hundred acres of land owned with a specific value) and to their sons over eighteen.[26] This was in response to what the elite saw as idle and disorderly people killing deer when they should have been working and taking care of their families. By 1771, game like bear, foxes, and wolves had pretty much disappeared in all of New Jersey south of the Northern Highlands, while in one Northern Highland Township, a five-year moratorium on killing deer was imposed in 1772.

By the mid-to-late nineteenth century, more people began to worry about the indiscriminate killing of America's wildlife and pushed for rules and other forms of wildlife protection. It wasn't, though, until groups of hunters formed clubs like the New York Sportsman Club (1844), Boone and Crockett Club (1887), and later the Audubon Society (1904), the American Bison Society (1905), Ducks Unlimited (1937), the Rocky Mountain Elk Foundation (1984), and more recently the Muley Fanatic Foundation (2012) that rules were put in place to protect these animals and increase their habitat for sport hunters. The Boone and Crockett Club was instrumental in producing what is now known as the North American Model of Wildlife Conservation that included a set of rules for hunting. They also created guidelines for trophy hunting that targeted the largest of the species (as large horns/antlers suggested an older male) so that the young breeding males would remain in the population. The idea was an effort at sustainability that, like the rules in the North American Model, was intended to make hunters more accountable for the number and age of species they killed, as well as for how they killed the animal. Neither the conservation model nor the guidelines for trophy hunting have withstood the test of time particularly well. As with grizzly bears and the Old West attitudes of subdue and conquer nature to lay claim to established notions of masculinity, after a many-year scientific study and millions of tax dollars spent to return the gray wolf to YNP in 1995, politically powerful hunters and ranchers in Idaho and Montana won legislative battles to turn the 2021–2022 wolf-hunting season into a wolf-killing free-for-all. That approach violates both the North American Model of Wildlife Conservation and the guidelines associated with trophy hunting.

To their credit, these hunting organizations have played a critical role in making coexistence palatable to those typically less interested in conservation. They have contributed and supported land acquisition and reintroduction efforts important to species survival, but for each club or organization, the end game is typically to protect their preferred species for sport hunters.

The Rocky Mountain Elk Foundation (RMEF), an organization that has enhanced elk habitat in the West and reestablished elk populations in many places east of the Mississippi, is still an elk hunting organization. With hunting elk as their goal, the RMEF supported the 2022 effort at delisting grizzly bears so they would no longer be protected under the ESA, while in 2020, the organization had taken the same position with wolves. To them, grizzlies and wolves compete with hunters for the elk resource. Charlie Booher, writing in RMEF's Hunting is Conservation section in their journal, the *Bugle*, highlighted this conundrum when he described how early conservation efforts were funded by hunting and fishing licenses that relied on "selling the opportunity to pursue "excess" wildlife."[27] By its design, this funding method prioritizes the conservation of *game* animals over those that are unable to contribute to the system in a financially measurable way. It becomes a rob Peter to pay Paul approach that can result in elk numbers exceeding the land's carrying capacity, all while non-huntable bird species disappear from the sky. To Booher, being a conservationist requires more than just buying a license but should also include a committed engagement to maintain a world that supports both our wildlife and our wild spaces.

By including hunting is conservation in its branding, RMEF gives a nod to conservation and in turn coexistence in the minds of people who typically can be opposed, just as including conservation as part of the organization's mission statement can enhance public perceptions of hunting. To get support from ranchers and farmers who do not want elk competing with their livestock for grass, increasing their tolerance for reintroduced elk by supporting predator removal and control might have been a compromise RMEF had to make. Although efforts like these can make coexistence and conservation fall along a spectrum of species acceptance, their goals still make predator and other "nuisance" species control a popular negotiating tool.

A person whose identity is tied to dominance over nature can help explain why predators generate such hostility. If a rancher loses a calf to weather, it is understood to be out of his control, but if he loses one to a wolf or grizzly bear, the loss has the appearance of being avoidable and suggests the rancher is less skillful at protecting his herd. For some ranchers, a high public regard is linked to their success at managing predators. To Robert Chaney in his work on grizzly bears, the hunter also plays a role in maintaining this dominance-over-nature structure. By killing a grizzly, the hunter has become more powerful than the bear. In an article on the importance of using bear spray to protect both humans and grizzly bears, Todd Wilkinson wrote, "While it may indeed be macho to sit in a saloon and speak with bluster about gratuitously blowing a grizzly bear away with a rifle or sidearm, it's dangerous, risky, and can be illegal if proved unjustified."[28] One victim of a grizzly bear attack who was hiking in a known grizzly bear area and chose to ignore ranger recommendations for safety by hiking alone without bear spray had told a campground concessions worker the day before his death that he was a "grizzly bear expert." His "expertise" led not only to his death, but to the death of the sow grizzly that had attacked him and the removal of her two cubs to the Grizzly and Wolf Discovery Center in West Yellowstone.[29] In another instance, as we got our own bear spray out at a trailhead for a hike in YNP, a man nearby made a point to show us his pistol and to say he would use it "if he had to." Since it is illegal to use a gun in the park, although you may carry one if the state the park is in allows it, why bring a gun on a busy hiking trail when bear spray is the legal and more appropriate option.

Claims to power like these maintain a man's dominance over nature and are played out in how men document the bloody battles of predator-prey relationships with trophies, stories, and photos of the violence. That documentation takes the viewer into what Chaney calls a "macho cul-de-sac of nature porn,"[30] where the idea of violent nature as porn complements the dominance and aggression of the "macho man."[31] This attitude of male power and hunting has been used to explain how the majestic wildlife found in America, Africa, and India, which includes bison, moose, bears, elk, elephants, and tigers, offers the hunter a high social value that in the animal's death can become his trophy. According to Jack Ashby in his work on Australian wildlife, "Moustachioed toffs in pith helmets traveled to [parts of the former British empire] specifically to murder mammals, in order to nail them to their walls as supposed demonstrations of their epic manliness." For him, unlike the high social value animals in America, Africa, and Asia have had, British perceptions of Australian wildlife as inferior and stupid removed the kangaroo or wombat from consideration as a manly trophy.[32]

Our relationship with grizzly bears is also tied to myth and metaphor. The myth of the bear's bloodthirstiness and hostility toward humans, the large ferocious bear metaphors associated with the grizzly's Latin name *Ursus arctos horribilis,* and the "need" to subdue an animal that, before the arrival of settlers with guns, "wandered boldly over his domain, master of all he surveyed"[33] allowed for their eradication from the plains and their removal from most human-occupied spaces. A bear that was called the "Monarch of the Wilderness" or the "Monarch of the Forest" was diminished in both number and form to live outside ever-expanding human boundaries. As part of the ideas associated with the productive use of natural resources, the grizzly bear was not only killed but captured for human entertainment. In the mid-nineteenth century, it was not unusual to find captive bears caged or chained at railway stations to perform tricks for or eat sweets from waiting passengers. As commodities, a "wild" grizzly could be sold for $15.50, while a trained one for $20.50. William Randolph Hearst saw the bear's economic potential through the lens of selling his newspapers. To achieve that goal, he arranged for the capture of an ever-more-rare grizzly to promote his paper's readership by exploiting the animal's impending extinction in California.[34]

By adding gender to the mix, the New West approach further threatens the masculinity of the Old West. The Old West's belief that masculinity occurred when a man subdued and conquered nature doesn't work well with the New West's attitude toward coexisting with nature. As part of a multi-member ranch association in an area adjacent to YNP in Montana, Malou Anderson-Ramirez and her family work to share the land with all manner of wildlife and rely on regenerative agricultural practices to maintain what they now see as their ranch's primary business, which is making and maintaining its critical soil resource. To maximize the land's potential, the Andersons rely on female range riders who, they believe, maintain a thoughtful presence on the land and bring with them feminine energy. According to Anderson-Ramirez, her best range riders are women. Another family member supported that by saying livestock need to be managed with sensitivity and respect and that honoring what is on the land allows for a cooperative relationship with nature to develop. The calming presence of female range riders helps cattle respond better in the face of predators, while a heavy-handed management approach can destroy cattle confidence. According to Anderson-Ramirez, confident cattle don't run from predators but, instead, stay together as a herd. By relying on this kind of energy rather than a more aggressive, masculine energy, sharing the landscape with predators can become a reality for more ranchers. For Anderson-Ramirez, those ranchers who do not see the land as sacred fail to exist on the land in a non-threatening way and that failure makes "[them] mad at the bears and wolves. [It also] adds to the embedded anger and resentment that exists over weather and even magpies, developing into an adversarial relationship with nature." The failure to maintain their land's high productivity requires an antagonist, and for the rancher that antagonist can easily become the wolf or grizzly bear (see figure 5.1).

An April 2022 Letter from Idaho in *The New Yorker* titled "Killing Wolves to Own the Libs?" reminds us of another set of competing forces when it comes to choosing to live with nature or to dominate it. That force is tied up with our identity, geography, and heritage but is also expressed through an individual's position on the political spectrum. A "blue state's" contingent of ardent supporters of grizzly bears and wolves now competes with a "red state's" antipathy toward those same animals, just as state's rights compete with the rights of the federal government. To predator fans, returning management of wolves and grizzly bears to the states is a recipe for failure since many states have demonstrated they cannot be trusted with managing species

Figure 5.1 Working Ranch Sign, 2022. *Source*: Barbara K. Jones.

they dislike. A common "blue state" feeling is that by relying on politics and power to make wildlife decisions, states in the Intermountain West have consistently failed to manage their predator populations according to science. That choice has resulted in successful lawsuits and threats about relisting affected species. In Idaho and Montana, the political decision to eradicate 85 percent of their wolves defied science but delighted some hunters and ranchers. However, the result of this eradication effort was the loss in YNP of nearly the entire Phantom Wolf Pack to wolf hunters operating along the park's northern border, while in the basin, where the Anderson ranch is, the wolves were, according to one family member, effectively wiped out. For the last two years, they have had ninety cows and calves out in "rough country" with old wolf dens and no loss of livestock. Yet today, "with all the savage hunters out there, the wolves are gone."

For many of those who live in these states, the federal government is viewed with suspicion, while the resentment Woodard identified as part of the Far West Cultural Region, along with a belief that the "libs" are ignorant "snowflakes" who know little about what life in the Rocky Mountain region is really like, has encouraged state officials and agencies to manage their wildlife resources as they see fit, not as if they are public trust resources. Yet wildlife on federal lands belongs to all of us, regardless of the "color" of the state (see figure 5.2).

Figure 5.2 Wolves in Yellowstone National Park. *Source*: Barbara K. Jones.

In many parts of the United States today, there is a shift away from these dichotomous views that have defined our relationship with land and wild nature. For hunting, its declining appeal, where only ten percent of Americans hunt, has led hunting groups and weapon makers to reach out to women and children to broaden their audience. Major gun dealers now sell guns in pink and purple, while taxidermists offer European mounts where the synthetic skulls can be colored in country-cottage hues. This effort at recruitment could change the way hunting is perceived and shift the practice away from its dominion-kill-what-can-be-killed narrative. The presence of twenty-first-century young people with their twenty-first-century sensibilities, where 71 percent queried in a 2018 poll considered hunting animals for sport to be "morally wrong,"[35] could make hunting less about dominion and more about coexistence, while the addition of women could make it less about power and more about being in nature. It could also move the imagery associated with hunting away from trophy shoots that include pictures of weapons, dead animals, and sex-linked organs like horns and antlers to what many hunters claim really matters, which is experiencing nature and providing food for their families. It could also mean changing the messaging of magazines like *Predator Xtreme* that offers information on legal wolf hunts yet accepts that a lot of people manage their resident wolf populations with a "shoot, shovel, and shut up" strategy, an approach that generates tremendous concern for the wolf's many fans. If hunters, as Mattson wrote in his *Worldviews and the*

Ethos of State Wildlife Management essay, truly have a "benevolent view of wildlife," a legitimate question to ask is why that more compassionate view is not typically expressed in hunting-oriented publications,[36] where wild nature, family scenes, and cooking could readily replace the images of big guns, dead trophy animals, and brawny men.

The hunters, ranchers, and farmers who have dominated the political landscape in states like Idaho and Montana but represent only a fraction of the American population will eventually overplay their hands. In the winter of 2021–2022, the loud public pushback against the politically driven wolf reduction decision in the states of Montana and Idaho had made relisting the gray wolf again a real possibility. The logic for this mass killing was based on the primacy of ranchers and hunters whose stories about wolves as shoplifters and killers or that a good wolf is a dead wolf continued to perpetuate a deeply held anti-wolf narrative. Their stories spoke of wolves as vermin and evil beasts rather than as wild animals critical to a healthy landscape, and the resulting rule change reflected the power of that messaging. In response to the poor optics around such aggressive anti-predator behaviors, the non-hunting/ranching public was encouraged to question the ability of state governments to separate politics from science when they claim to manage species within their borders appropriately. Public concern over wildlife has also encouraged those agencies in charge of managing the bison that leave YNP in winter to move more of them to tribal lands and other preserves rather than have them killed in Beattie Gulch. Changes like these suggest the dominion-over-nature metaphor is being outcompeted by one of coexistence and that the narrative of the New West, where protecting nature can be called a masculine experience, is positioned to take precedence over that of the Old West, where masculinity is defined by conquest. Although a significant and persistant problem in the states of the Intermountain West is getting those state legislators and governors to recognize and accept this changing reality.

THE FORCES OF MYTH, TRADITIONAL KNOWLEDGE, AND MODERN SCIENCE

Mary Roach, in her book, *Fuzz,* wrote that her idea of what an elephant was came from reading about one named Babar that wore white spats and a bright green suit. That Babar imagery created in her an understanding of elephants as slow-moving and gentle animals that generated nothing to fear.[37] Yet in places where elephants live, the reality is far different. An unexpected encounter with a wild elephant, elk, or bison can be extremely dangerous, especially when a bull elephant or elk gets an extra jolt of testosterone during the mating season. The recognition that wildlife can be unpredictable and dangerous has shifted how parks with wildlife educate their visitors. Instead of emphasizing the threat predators like bears and wolves can pose, YNP calls all wild animals dangerous. Just like a testy bull elephant, a cranky elk or bison in rut can do serious damage to the human body or someone's property. This new approach has begun to contextualize all species in a wild setting so that the grizzly bear is one kind of dangerous and the bison another. While I was at YNP's wolf acclimation pens that were used during the wolf reintroduction in 1995, we came upon a bison resting nearby. Although we were outside the park's required twenty-five yards from a bison, it was interesting to hear my guide say, as we moved to leave the pens, that we needed to make every effort to avoid the bison because they can be one of the most dangerous animals in the park. Rethinking myths helps us better understand that grizzly bears are not typically man-eaters, wolves are not indiscriminate vicious killers, and bison aren't fluffy cows to be petted. Instead, they are all wild animals that can behave in unexpected ways.

The opportunity to understand wildlife in a more holistic way comes through an integrative approach called the consilience model. This model looks at the convergence of evidence from unrelated sources that might include traditional ecological knowledge (TEK) and modern science, wildlife viewers and YNP wolf biologists, ecotour operators and ornithologists, or hunters and state fish and wildlife managers to develop new ways of knowing. The goal of this approach is to rely on evidence from independent, unrelated sources to see where they "converge." That convergence of information can produce strong conclusions and, due to the variety of sources, offer new ways of thinking about information. To Wyoming's Game and Fish large carnivore specialist, Dan Thompson, affirming wildlife success stories with peer-reviewed and published data allows for a "synergy of story with science" and a better reality regarding species recovery and resurgence.[38] Anthropologists also rely on this approach when writing ethnographies. Interviews, archival work, and observations can produce a concordance that highlights important shared themes within a community of competing stakeholders. Anthropologist Clifford Geertz called this approach thick description, while environmental anthropologist Elizabeth Van Dolah of the Nature Conservancy relied on what she saw as a form of "deep hanging out" to make sense of the social and environmental challenges people living on Maryland's Eastern Shore face.[39]

An interesting example of the consilience model in action occurred with the sea wolves of the Great Bear Rainforest in British Columbia. It wasn't until the 1930s that these coastal wolves were determined to be unique and were "discovered" by a zoologist named Ian McTaggart-Cowan; yet, according to coastal First Nations people, the sea wolf didn't need to be discovered because it was already known and embedded in their origin stories.[40] The blend of traditional knowledge from First Nations people's awareness of sea wolf behavior and the technology associated with science that allowed for genetic studies through fecal sampling has offered important insights into how the sea wolf lives. The wolves' feces have provided information on hormone levels, parasites, and bacteria, as well as information useful for location mapping, species' range, and habitat, while TEK has offered insights into sea wolf eating habits. This mix of information is giving scientists and First Nations people an opportunity to better understand the complexity of this animal. That knowledge in turn offers better ways of identifying threats to the species as well as ways to mitigate those threats. For Churchill Wild's wolf program in the Hudson Bay Lowlands of Canada, the integration of scientists, citizen scientists, and the Cree people who pair traditional fieldwork with Indigenous oral history has made this program unique in increasing our understanding of wolves living in this region of North America.

For some Indigenous communities, the domination of western science over TEK is not always well received or even appreciated. To Indigenous Arctic communities, the polar bear, a species that has maintained the Arctic peoples' bodies and souls, is understood to be the King of the Arctic.[41] This label forges a way of seeing polar bears not as a wild animal outside the human experience but as a member of a shared social community. Flying over bears, or capturing, tranquilizing, and tagging them can conflict with the tenets of TEK that rely on respect, responsibility, and reciprocity when dealing with nature. For Inuit people, their traditional ecological knowledge comes from an intimate connection established from a long association with Arctic wildlife, so it offers other explanations to polar bear management that do not necessarily mesh with science. They see TEK as something not necessarily rooted just in lived experience and cultural understanding, but instead on empirical evidence. An Inuk hunter, Gabriel Nirlungayuk, spoke of how their knowledge is always evolving to match changing environmental conditions because to an Inuit living in the Arctic, "if your life depended on [safe passage], you'd better know what the ice condition is."[42]

In the *Species Status Report Polar Bear in the Northwest Territories*, the Kitikmeot Regional Wildlife Board wrote how ". . . western science and modelling ha[ve] become too dominant." These feelings were documented in the report's reasons for the assessment summary of Indigenous Community Knowledge (ICK) and Scientific Knowledge (SK) regarding the polar bear. Documented as important additional factors to this assessment were concerns about invasive research techniques that Indigenous community members felt were negatively impacting polar bear health. According to the report, "Harvesters and elders from numerous communities have discussed how chasing and immobilizing polar bears with helicopters so that they can be tagged can 'spook' bears."[43] Yet to scientists these techniques are essential to assessing bear health and their use, according to them, has resulted in no proven negative effects. In another instance, observations by Indigenous people identified bears eating the entire seal instead of just the blubber or on Svalbard, a

Norwegian archipelago north of the Arctic Circle, residents have seen polar bears eat reindeer when seals are not available. For Indigenous people, this change is the result of nutritional stress due to rising polar bear numbers that are also pushing polar bears into human spaces where additional food sources are more readily accessible to them. Their presence in unexpected human places is also leading to more human-polar bear conflicts, and with the recent losses of life to polar bear attacks, there is a feeling the bears may have exceeded the coexistence threshold of the Nunavummiut (residents of Nunavut in northern Canada).

Conflicting with this view are national and international environmental and non-governmental organizations, climate change advocates, and the global community of polar bear enthusiasts that have dominated the discussion on the status of polar bear populations. Polar bears have been the face of conservation campaigns and the iconic symbol of climate change such that TEK is competing against a firmly entrenched idea about polar bear survival as sea ice conditions change. Scientists do not believe polar bear numbers are increasing, but rather that climate change is forcing polar bears into towns and residential areas. To them, changing sea ice conditions have led to a decline in the body condition of seals and subsequently the polar bear's body health. Conversations with residents of Svalbard highlight how they see themselves as occupants in the polar bear's place. It is illegal to track them, to locate them for visitors, and no one is allowed to leave the settlements without a permitted, and armed escort who is trained to minimize polar bear-human confrontations. Coexistence is so highly valued on Svalbard that people are required to adapt to nature. Residents and visitors cannot drive snowmobiles on glaciers or frozen water because it disturbs the polar bears when they are trying to catch seals at breathing holes on the ice. This is taken so seriously that the territorial governor polices the glaciers and if "nature's rules" are broken, the fine is 5000 NOK or about $470 and a yellow card warning. If the individual's behavior is too egregious, the violator can be forced to leave the island. Unlike in the United States where coexisting with top predators like the grizzly bear is not expected and there are typically few consequences for killing one, on Svalbard the needs of the bear are the priority and people are obligated to adapt to the bear's needs. For those who kill a polar bear, the consequences can be severe. By relying on multiple points of view that include TEK/ICK and SK, new ways of knowing and achieving coexistence become possible. This blending of knowledge results in a more comprehensive understanding of the social and ecological realities of a species like the polar bear and how we can better coexist with them.

Unlike TEK, settler's knowledge was formed by the language of subduing and conquering that resulted in an ecological awareness based on domination over nature. That relationship helped to organize how the academic field of fisheries and wildlife management was designed, with its roots in the academic disciplines of agriculture and forestry. Since both agriculture and forestry rely on the principles of producing and harvesting the greatest sustainable yield, relying on that paradigm for wildlife results in what Charlie Booher called a system of management rather than a system of conservation. In a 1948 academic *Wildlife Management* book, the author Reuben Edwin Trippensee wrote how "wildlife management has a sound economic basis," where "crops of wildlife can be grown on land not suited to other crops."[44] His book series also included topics like upland game, fur bearers, waterfowl, and game fishes, suggesting the language of crops and game was used to encourage wildlife management students to see wildlife as a commodity. In his 1987 book *The History of Wildlife Management in Wyoming*, Neal Blair wrote how a wildlife management program's success was measured not by how much good was done for a species, but rather by the number of animals that were harvested. This was supported by the many harvest reports he included in his book.[45] Both texts give insight into how future wildlife managers were taught to see wildlife.

The language of production and harvestable yields conflicts with the language of conservation and its endgame of foresighted use, preservation, and renewal of our natural resources for the "greatest good of the greatest number for the longest time."[46] That conflict reminds us that extracting resources from the land needs to occur with conservation in mind; otherwise, today's resource users will have traded the health and prosperity of the future for the needs and wants of today. The nation's first conservationist, Gifford Pinchot, had his conservation moment when he understood that nature was the result of interlocking relationships working together as parts of a single whole. From that, he realized how we used nature was the problem that needed to be addressed. By communicating this idea about our use of nature to President Theodore Roosevelt, conservation soon became a more familiar word to Americans.[47] Conservation as a way of living with nature has encouraged more people to move from seeing wildlife as a commodity, where wildlife is managed through a top-down system of production designed for specific users, to a future-oriented, renewal-based approach that preserves nature for all of us for the longest amount of time.

Whether we see wildlife management through the lens of production or through the lens of renewal determines how we relate to wildlife. Those who see nature through the lens of production tend to be traditionalists or hold domination values, while those who see nature and wildlife as part of an interconnected whole are understood to be mutualists. Michael J. Manfredo et al., in their work on the human dimensions of natural resources, used those labels to explain wildlife's social value shift that occurred with the increasing urbanization and rapid economic growth that occurred post-World War II. These changes encouraged people to transition from traditional or domination values to values rooted in mutualism when they considered human-wildlife relationships. Their work identified domination as an effect of the Protestant Reformation, where natural resources were intended for productive use. Mutualism is a view that sees us more as members in a shared human-wildlife social community, a community Indigenous people today would say has long been part of their value system. These more mutualistic values emerged as we domesticated the environment and removed people from the potential dangers of wildlife; yet by isolating ourselves from wildlife, we reduced our social connections to other species. That isolation also led us to assign nonhuman species human attributes and the moral consideration typically assigned to others of our own species, just as it encouraged a new generation of social scientists to incorporate the point of view of wildlife into their multi-species ethnographic research. By exploring and identifying the social and ecological needs of wild animals, we not only can improve our connectiveness to these species, but also learn to assign them traits more appropriate to their wildness.

To Joseph Rump of the Voyageurs Wolf Project, in a talk called "Wolf Personalities: Ecosystem Impacts and the Possibility of Wolf Cultures," this understanding of what it means to be a wild wolf comes in part from recognizing unique individual wolf behaviors or personalities as a prerequisite for developing wolf culture. To make the case for animal culture and its connection to unique animal behaviors and personalities Rump considered the stories written about very specific wolves by Rick McIntyre, a long-term wolf observer and author of many books about wolves, including *The Rise of Wolf 8: Witnessing the Triumph of Yellowstone's Underdog* and *The Alpha Female Wolf: The Fierce Legacy of Yellowstone's 06*. By highlighting the lives of specific wolves, it then becomes possible to better understand more holistically the species in its wild context.

The results of Manfredo et al.'s research have applications for grizzly bears, gray wolves, and even mountain lions. The states of Idaho, Montana, and Wyoming are home to nearly all the remaining grizzly bears in the lower forty-eight states, yet they also have some of the highest numbers of people with traditional or dominant value systems. For Montana, 38.5 percent of the residents surveyed were traditionalists, with 24.5 percent identifying as mutualists. In Wyoming, 47.8 percent were traditionalists (there was a four percent increase between 2004 and 2018), with 21.7 percent mutualists (with a 3.3 percent increase over the same years), while in Idaho, 38.1 percent identified as traditionalists and 23.8 percent as mutualists. Of the three states, only Wyoming saw an increase in traditionalist views over the fourteen-year study period.[48] These numbers are interesting, but the irony is that grizzly bears and gray wolves live in states where, to a group of powerful people, they are not particularly welcome or popular. Gathering social behavior information like this is critical to consider when decisions regarding who should manage public trust resources are made.

This enduring conflict between economics, ecology, and social tolerance that makes such compromises hard to achieve was on display recently when decisions regarding how the wolf was to be managed in Montana, Idaho, and Wyoming were being made. The pressure from many in the ranching and hunting communities encouraged the state governments in Montana and Idaho to give hunters the opportunity to reduce wolf numbers from 1,100 in Montana and 1,500 in Idaho to what political leaders called a sustainable number that was never scientifically determined. For wolves, their successful recovery conflicts with a cultural heritage that for generations has seen wolves as the enemy. Although the economics of wolf watching has done little to change this perspective, the disastrous flood of June 13, 2022, that decimated the economies of the gateway communities surrounding the north and northeastern parts of YNP in Montana might have. The flood made Montana officials very uneasy about generating controversies that could potentially damage Montana's outdoor economy. This is significant since wolf tourism is an important economic driver in the Yellowstone region, with fifty percent of visitors hiring a guide to watch wolves, while only fourpercent do so for hunting trips.[49]

For wolf advocates, the hope was that returning to a less controversial one-wolf hunting quota in each wolf management unit from a proposed ten wolves in a newly combined wolf management unit adjacent to the park would reduce negative media attention while protecting a critical ecological and economic asset. This is of particular importance because a high number of park wolves were killed in two wolf management units directly north of the park near Cooke City, Montana in the 2021–2022 wolf-hunting season. One park ranger believed it was only four or five hunters responsible for killing nineteen Yellowstone wolves in the Montana Wolf Management Areas adjacent to the northeast section of the park. The negative optics from this perceived assault on wolves made many people question the state's ability to manage their wildlife resources through science rather than culture. A similar concern is occurring in the states of Idaho, Wyoming, and Montana with grizzly bears, where the states' governments and wildlife agencies are working to get the grizzly bear delisted so its management can return to state control. Chris Servheen, a retired bear biologist for the USFWS and once a prominent advocate for delisting the grizzly bear and returning its management to the states, has turned entirely away from that idea because of politics and "a wave of fact-free 'hysteria' sweeping the Rocky Mountain West."[50] Before his retirement in 2016, he felt wolves and bears were in a good place, but today he thinks Montana is managing its wildlife as a "clown car of absurdities." To him, the people in charge of managing wildlife are politicians who are totally misinformed about what is really going on, so they make decisions based on ignorance and their "crazy feelings about 'I hate predators.'"[51] At a May 31, 2024, public conversation sponsored by *Mountain Journal* called "Heating Up: Grizzly, Wolf Management, and Climate Change," Servheen spoke about how state legislators and governors have been passing legislation that implements "harmful anti-predator practices." To him, by demonizing predators, these states justify the unnecessary killing of these animals. An attitude about wolves and grizzly bears that sees them through dominance-over-nature metaphors plays into the Old West attitudes of masculinity and power and flies in the face of the New West that sees conservation as a more masculine and powerful alternative. In this reality, predators are serving as proxies in a battle between Old West and New West values.

With the election of the first Republican governor in Montana in over fifteen years, Greg Gianforte, a wealthy hunter who sees wolves as his nemesis, harnessed the power of his office to engage in "payback" for what he and some of his fellow hunters see as the wolf's misfortune at surviving and thriving in Montana. Governor Gianforte, along with a Republican majority in the state legislature and agencies packed with Gianforte appointees, chose to manage wolves through the lens of culture rather than that of science. They removed all wolf-hunting quotas to achieve what Ryan Devereaux, writing for *High Country News*[52] called turbocharging the efforts of making wolves fair game for hunters and trappers. Since no one really knew how many wolves there were in Montana, the decision by state politicians to kill four-hundred-fifty wolves so the population could drop to a "sustainable" level was based on nothing but whim and spurious claims. By making decisions outside of science and without an accurate count of wolf numbers or how many were necessary to maintain sustainable packs, wolfers got to decide for every other Montanan what was an acceptable loss of wolves in the state. Not only did they make the kill goal high, but state officials, by extending the hunting season, giving individual hunters and trappers licenses to kill twenty wolves each, allowing the use of bait to lure wolves off protected lands, permitting aerial hunting, and night hunting with bright lights, night vision equipment, and snares that frequently kill other non-targeted wildlife in addition to people's pets, made no effort to encourage fair chase. This was a politically motivated wolf kill that resulted in public outcry, a loss of tourist revenue to communities adjacent to YNP, and the extermination of an entire highly studied and viewed YNP wolf pack. Carter Neimeyer, a former federal trapper who worked for Wildlife Services as their top wolf official, called the wolf kill "payback now, clear and simple." He went on to say that "it's just a bunch of politicians reinforcing a bunch of mercenaries, who are having a grand ol' time killing wolves."[53]

In 1995, when the wolves were returned to YNP, the park service had to hide armed guards in blinds near the wolf acclimation pens. People threatened to poison the wolves, while Wyoming Republicans passed a measure offering $500 to anyone who killed a wolf outside the park. The state was even willing to pay the wolf hunter's legal fees if they were charged under the auspices of the ESA. Hunters thought a way to get rid of the wolves might be through a method tried in Minnesota called the "shoot, shovel, and shut up" approach. This method to some degree appealed to people like Chad McKittrick, a Red Lodge, Montana wolf hunter, who, with little remorse, killed, then skinned and beheaded a wolf that had left the park. He was arrested because he apparently ignored the "shut up" part of the approach. To demonstrate his disdain for a system that returned the wolf to western lands, he rode his horse through Red Lodge in its Independence Day parade, with a pistol at his side and a T-shirt on his back that had the words, "Northern Rockies Wolf Reduction Project" on it. To Devereaux, this kind of behavior was part of what he called a "state-sanctioned outburst of revanchist political aggression and revelry."[54] The feeling by many left in agencies working to manage Montana's wildlife was that the legislature and the Republican Party wanted "revenge" for the 1995 wolf reintroduction. Payback or revenge as a management tool is rooted in cultural values, not in science or with any consideration for others whose value system might not mirror your own. It also is reminiscent of both Baumeister's words about wildlife management strategies that teach people to kill rather than to hunt and Audubon's pal who built pits to capture and torture wolves who he believed had killed his sheep and colts.

In 2020, killing wolves in Montana became part of some hunting outfitters' campaign messaging for Donald Trump. To these outfitters, the presence of wolves on the land was an example of "liberals decid[ing] what's best [for Montana's] land and resources."[55] Once the wolf-hunting laws were relaxed with the election of Gianforte, well-known anti-wolf families in the area north of YNP promoted their wolf kill successes with shared photos on social media that encouraged commentors to express a "kill 'em all" mentality.[56] Their joy came from what Marc Cooke, president of the nonprofit organization Wolves of the Rockies, believed was the "pleasure [they got] in killing that wolf because it hurt and cause[d] pain [to those] in the Yellowstone wolf-watching community."[57] By politicizing the wolf, a dead wolf became a mascot for those who saw the wolf's return as government overreach and has made increasing the social tolerance for wolves in the Intermountain West a challenge that might take generations to overcome.

The wolf is only one example of the many species that, through successful reintroductions or habitat improvement, is challenging the relationship between humans and wildlife in unexpected ways. In the case of the wolf reintroduction, the ecological and economic boon to the region has been well documented. Wolf tourism has generated new business opportunities that have rippled through Park County, Montana. This revenue keeps communities like Gardiner and Cooke City, Montana, thriving throughout the spring, summer, and into the fall months. Yet within these same communities, those driven by wolf hatred played a significant role in the losses YNP's wolves sustained during the 2021–2022 wolf hunt. Out of the twenty-five park wolves killed, more than half were females; at least four were pups, and seven were yearlings.[58] Fifteen individuals were responsible for killing twenty-one of the wolves that had left the protection of the park, and of those fifteen, four individuals were responsible for killing more than ten. According to Devereaux in his research on the hunt, one local who lives in Jardine, a small community adjacent to Gardiner, had a front-row seat to the wolf fight. This individual watched men from Gardiner, Emigrant, and Livingston, Montana, engage in something that he claimed was "not a wolf hunt. It's killing is what it [was]."[59] The hunters used cell phones to coordinate a kill—something that is illegal to do in Montana and violates the ethic of fair chase—and they attempted to lure wolves out of the park, where they skinned wolf corpses and left their bodies in the snow. To these men, the wolves were the enemy, and by blaming the decline in elk numbers on the wolf's presence, they were justified in killing them in any way possible. Contrary to this perceived reality, the number of elk in Montana exceeds the land's carrying capacity in much of the elk habitat, such that cattle and sheep ranchers in Paradise Valley (the valley north of Gardiner, the northern entrance to YNP) are subsidizing the elk's natural diet with grass from their irrigated fields. Ironically, the inability to increase social tolerance for wolves has left wolves out of any elk management conversations, so elk are managed in a counter-intuitive way.

Since current elk management strategies have offered little to make anyone happy, Henry Worsech, a seventeen-year veteran of FWP, came out of retirement to lead the FWP in finding a new approach to balancing landowner concerns over high elk numbers with hunter opportunity. Since his appointment, Worsech's efforts at shifting how elk are managed have resulted in a lawsuit, some harsh editorials, a petition asking public officials to uphold the public trust doctrine of wildlife management, and a Rural Montana Foundation report that accused the FWP of "disregarding landowner concerns and being more responsive to politics than science."[60] At issue were public land elk hunters who complained that hunting on state and federal land was a fruitless effort, large landowners who were angry they had to absorb the economic losses excessive elk numbers put on their forage and fences when elk moved to private land to escape hunting pressure, while outfitters were upset that the lucrative bull elk hunts were out of their reach because of the uncertainty of the state's hunting tag draw system, and wildlife viewers argued that the state appeared to be moving toward a privatization of wildlife through a landowner incentive program that has been called "bulls for billionaires."[61] The difficulty in managing elk in Montana is exacerbated by how FWP manages wolves. Because elk are under a lot of hunting pressure on public land, they tend to congregate during hunting season on private land out of reach of many hunters. Since wolves don't recognize public and private land boundaries, by moving across and through these bounded spaces in search of prey, wolves would have kept the elk moving too. The presence of wolves would give elk hunters more opportunities to use their elk tags, would reduce the expense to private landowners of maintaining large elk herds, outfitters would have more predictable access to bull elk hunts, and wildlife advocates would be reassured that elk were treated as a wild public trust resource. Instead, by not including wolves in the elk management plan, these issues have not been adequately addressed, forcing Worsech to devise another elk management plan in 2022. The failure to co-manage elk and wolves effectively suggests these species in Montana, like the white-tailed deer in Maine and Minnesota, are being managed as political entities rather than as biological ones.

In February 2022, when Montana's Region 3's, a large wolf management area adjacent to YNP, wolf-kill quota was reached at eighty-two—with an additional three killed during the post-hunt grace period—that total included over 20 percent of the entire park's population. To understand the significance of that number, in the years following the 1995 reintroduction, on average, two to four park wolves were killed in land adjacent to the park. There was an outlier in 2012 when twelve wolves were killed after the hunting quotas temporarily lapsed, but the rationale for the 2021–2022 wolf-hunting season was based on no reasonable explanation. Yellowstone superintendent, Cam Sholly, wrote a letter to Governor Gianforte asking him to halt the killing of wolves in hunting districts adjacent to the park. He reminded the governor of the long-standing cooperation between Yellowstone and the state of Montana on game management issues, such as brucellosis in bison, and that the elk population across the entire state was at or above management targets. He also reminded the governor that in the past three years, only once were wolves involved in livestock predation in Park County. He then reiterated the economic impact wildlife tourism brings to the region. Sholly wrote in the letter, "I am still unclear what state objectives are being met by lifting quotas on these two units." None of this moved the governor because he said he had no "authority" to act even though it was his administration and political will that changed the rules so that wolves became much easier to kill. In January 2022, Sholly forwarded his letter to the agency Gianforte said was in charge, which was the FWP. The letter didn't motivate the FWP to impose quotas along the park's northern boundary either, and the wolf kill continued until the 82-wolf allotment for Region 3 was reached.[62]

An additional problem with the wolf killing spree was the destabilization of established packs. By killing the lead female, a pack might not reproduce, it might have more than one breeding female, or younger wolves without a lead female might get into trouble. In addition, pack boundaries could shift, packs could dissolve, and pack survivability could become more challenging. According to Doug Smith, the now-retired park biologist who worked with the park's wolves since 1995, all this can occur because the pack's social fabric has been torn. The importance of an intact social fabric has been observed not only in wolves but with other species as well. Species that live as families maintain a social structure that encourages the young to operate within a specific set of rules. Once you remove that structure, the young of the species can go rogue.

Within the park, the wolves are fully protected, so their numbers naturally fluctuate based on the park's carrying capacity. Since park wolves live in a natural environment that is regulated by predator-prey relationships, mate availability, and weather, their numbers can change. There can be losses to the lead female or male, so pack membership and pack boundaries can shift. The food of choice can also shift through natural processes, but the intensity of the loss from anthropogenic forces can radically exacerbate that change. During the 2021–2022 hunt, at least four of the park's packs suffered tremendous losses, leading either to their dissolution or the inability to effectively hunt their prey of choice, or what some wolf watchers identified in the park as the unusual presence of two breeding females within one pack near Slough Creek in Lamar Valley. If wolf management doesn't change, the remaining park wolves can continue to live safely in the park, but simply by crossing a human-defined and invisible park boundary, they move from living within a wolf-safe zone to one that is based on the area's current low social carrying capacity for wolves.

Hunters have identified "pinch points" along the park's Montana-Wyoming border that funnel wolves along narrow known trails so the wolves, if they step over the boundary, can be shot by those hunters waiting on the "right" side of the boundary.[63] In Alan Kesselheim's article called "Yellowstone's Wolf Toll," he wrote about the invisible boundaries that exist at the edges of the park wolves' territories. Since those boundary areas maintain spaces that tolerate different social carrying capacities from wolves as vermin to be exterminated on sight to wolves as the highly valued wilderness ideal, to encourage greater coexistence, the figurative language of vermin must be reconciled with that of the wilderness ideal.

In a winter 2022 interview for *Wyoming Wildlife* between John Turner, who in 1989 was appointed to be director of the U.S. Fish and Wildlife Service (USFWS), and Wyoming Game and Fish Department wolf biologist Ken Mills about the reintroduction of wolves to the GYE, Turner addressed the challenges of finding common ground between the people in the Rocky Mountain West and those around the country interested in returning wolves to the GYE. Turner grew up on a ranch hunting, fishing, and guiding, so he knew the issues, but he also thought it would "be great to hear a wolf here in our backyard, here in Jackson Hole and Yellowstone."[64] His approach to achieving this goal was to make sure the states of Wyoming, Idaho, and Montana had both a voice and a management role in the wolf reintroduction process. As wolf recovery biologists for the USFWS, Turner's messengers were the face of this effort. Throughout the extensive hearing process, these biologists got their tires punctured, and they were spit on by angry attendees. Passions ran high in these meetings, but eventually a process was ironed out whereby once the target numbers were met, management of the wolves would return to the states. For Wyoming (along with Montana and Idaho), managing wolves has been a challenge state wildlife managers have been unable to effectively address; yet many of their difficulties could have been avoided if the people of Wyoming were offered a path to increasing social tolerance for wolves post-reintroduction. In Turner's interview, he suggested there was such a path that did stitch cultural heritage and wildness together. In a September 1971 *Wyoming Wildlife* article about bald eagles, he asked, "Can Wyoming do without its bald eagles?" He answered it himself by saying "probably so . . . but the loss would mean Wyoming had changed—that we have abused her land and lost that which has made our state unique in today's world. It will mean we as people and as a state lost our ability to sustain a very important ingredient of our heritage—that of wildness." When he was asked at the end of the 2022 interview if that answer could apply to wolves also, Turner said, "Hell, yes."[65] By saying that about wildness, Turner recognized the presence of wolves in Wyoming to be an important part of Wyoming's cultural heritage, and to lose wolves meant Wyoming was no longer a place that valued its land and its wildness. That message gets lost when wolves are seen as vermin.

To encourage a more mutualistic value system means to encourage multiple forms of coexistence. Even though the state governments and wildlife management agencies in Montana, Wyoming, and Idaho might have a less than welcoming attitude toward some of America's more iconic species, it does not mean there are no efforts to mitigate wildlife loss or to encourage wildlife recovery. In the summer of 2021, the Livingston, Montana Depot Museum sponsored an exhibit called *Invisible Boundaries: Exploring Yellowstone's Great Animal Migrations*. This exhibit explored the dual ideas that animals migrate, and humans draw lines on maps. These lines can be as obvious to an animal as an interstate or as obscure as a park boundary line. One grizzly bear, over a period of twenty-nine days in the fall of 2020 and twenty-four days in the spring of 2021, tried at least forty-six times to cross Interstate 90 in Montana. This was a bear that had left its natal range, and through this "natal dispersal" could, according to Montana's Fish, Wildlife, and Parks (FWP) bear biologist Cecily Costello, "provide the best hope for population connectivity."[66] Mapping migration routes have also provided information useful to landowners and agencies so that wildlife-friendly fences and crossings can become part of any management plan. Since pronghorns can't jump particularly high, in 2008, a local Wyoming land trust started retrofitting more than five hundred miles of fences along the pronghorn migration corridor. By raising the bottom strand of barbed wire, pronghorns can crawl under the fences along their migration routes. The state of Wyoming has made enhancing wildlife movement with improved wildlife corridors like these a priority.

The broad appeal of wildlife-friendly stories and metaphors is making it likely that more ranchers and hunters will begin to accept predators and other inconvenient species as compatible with their ways of living. Some ranchers in the Intermountain West already appreciate the price bonus their beef earns for raising cattle on predator-friendly landscapes. New product branding that highlights rancher efforts at coexistence or, more specifically, predator-friendly ranchland can compete in the marketplace like orangutan-friendly products or tuna caught in dolphin-free nets have already done. The Anderson family ranch is making a first effort at selling their beef through their unique wildlife-friendly label. They are going to introduce their beef during the fall grizzly bear viewing season when tourists travel twelve miles up into the basin where they live to watch grizzly bears eat caraway roots in what was, a brief time earlier, their summer cattle pasture. Ranchers are not the only ones trying to adapt to living with predators. Hunters in Montana and Idaho are also concerned that the change in wolf hunting rules has damaged hunters and hunting as a brand, while in Oregon, hunting organizations have joined with conservation groups to fund rewards leading to the conviction of wolf poachers in northeast Oregon. So far, over $80,000 in reward money has been raised. Members of hunting communities have worked hard to improve their brand, but by disregarding rules of fair chase and enthusiastically targeting wolves, a small number of hunters have tainted the image of hunting. Just like a few misbehaving wolves make it hard for every other wolf to survive in the wild, a few hunters or ranchers can damage the public's perception of them and their ways of life..

Accepting that living with nature can present its own challenges that those in urban areas don't witness daily, the coexistence umbrella must be welcoming to people with vastly different lived experiences. A rancher who has lost a sheepdog to a wolf will have feelings about wolves that a dog owner in Asbury Park, New Jersey, might never experience or understand. Those competing worldviews need to be recognized, and through that recognition, reasonable compromises that lead to enduring solutions can be made.

Conversations with people are essential for common ground to be found. Doug Smith at the "Heating Up" public conversation lightheartedly suggested that to find common ground with wolves in the Intermountain states, we need to find "a story that ties horses, guns, and wolves" together in a positive way.

As we move toward making the coexistence umbrella more inclusive, it becomes essential to remind ourselves of language's power in creating ways of seeing. For government entities like the Environmental Protection Agency (EPA), Animal and Plant Health Inspection Service (APHIS), the United States Department of Agriculture (USDA), and the Department of Health and Services, applying the pest, vermin, or livestock label to entire species is not useful. Since perceptions of nature are human constructs, the subjective choices we make in categorizing species can have far-reaching consequences. For coexistence to be successful, we need to begin to ask ourselves who gets to decide what makes a species vermin or livestock and how do those labels forge boundaries of separation between us and them that challenge their ability to survive.

NOTES

1. Wind River Wolf Management Plan 4.
2. Wind River Wolf Management Plan 2007: 4.
3. Lopez 112.
4. Lopez 112.
5. Lopez 113.
6. Lopez 119.
7. Coleman 2.
8. Lopez 139.
9. Norris 42.
10. Lopez 142.
11. Nash 27.
12. Lopez 142.
13. Coleman 5.
14. Coleman 11.
15. Wolf Wars: America's Campaign to Eradicate the Wolf.
16. Personal Memoirs Ulysses S. Grant.
17. The effect's name came from the novel *Beau Geste*, where the French troops propped up their dead, so their numbers looked larger to the enemy.
18. Brewster 169.
19. Coleman 12.
20. Stinson 7.
21. Stinson 7.
22. Chaney 16.
23. Mattson 30 (no. 5), Manfrado 40.

24. Mattson 21 (no. 4).
25. Braitman 86.
26. Wacker and Clemens 55.
27. Booher 38.
28. Wilkinson, October 29, 2017.
29. Repanshek. March 5, 2012.
30. Chaney 34.
31. Chaney 34.
32. Ashby 309.
33. Hanna 57.
34. Braitman 86.
35. The Economist/YouGov (2018).
36. Mattson 13, 25 (no. 3).
37. Roach 59.
38. Heinz, May 12, 2023.
39. Kane.
40. Tong 32.
41. 201 Cecco.
42. Cecco.
43. Northwest Territories Species at Risk Committee 5.
44. Trippensee V.
45. Gagliardi 15.
46. Pinchot 326.
47. Dekret.
48. Manfredo 4.
49. Park County Environmental Council. 2022 Wolf Trapping and Hunting Quota Talking Points.
50. Devereaux, January 29, 2023.
51. Devereaux, January 29, 2023.
52. Devereaux July 27, 2022.
53. Ibid.
54. Ibid.
55. Ibid.
56. Ibid
57. Ibid.
58. Kesselheim 19.
59. Devereaux.
60. Rural Montana Report 2.
61. Eggert. September 8, 2022.
62. Kesselheim 18.
63. Ibid 21.
64. Gocke and McWhirter, 40.
65. Ibid 44.
66. Pashby 20.

Chapter 6

Made Beaver

How Did a Beaver Become "Made?"

Today's North American beaver (*Castor canadensis*) traveled across the Bering Strait from Eurasia about seven to eight million years ago. It is one of only two remaining species in the Castor genus and the Castoridae family, with the other being the Eurasian beaver (*Castor fiber*). They, like mice, rats, guinea pigs, squirrels, and porcupines, form the Rodentia order. According to Lewis H. Morgan, writing in his 1868 tome on the American beaver, the beaver, because it lives on the "coarsest food," is slow of motion upon land, and has a brain without "those convolutions" that when present indicate mental power, and is a "low animal" that with its body design allows it to "rise no higher than the rat, the porcupine, or the squirrel."[1] For the beaver, being categorized with a rat in a human world is a tough place to be. Yet for all that, in his book called *The American Beaver and His Works,* Morgan took over three hundred pages to extoll the animal's sagacity and productive industry that raised the beaver in "human estimation" to a "very respectable position . . . for intelligence and architectural capacity."[2]

Such cognition and engineering skills allowed the beaver to spread in numbers estimated to be as high as four hundred million across most of the North American continent, from Alaska and Canada south to northern Mexico. Their numbers were so extensive that in 1852, as railroad tracks were being laid through what was then called wilderness toward Lake Superior in the Upper Peninsula of Michigan, the railroad passed through what Morgan called "beaver districts."[3] These districts were areas full of dams, fallen trees, lodges, streams, and beaver meadows, suggesting the animals were not confined to a park or pond but were part of miles and miles of beaver-engineered wetlands. In a beaver district, the beaver was not considered to be a nuisance or pest but instead an industrious landscape manager and as such not constrained behind human-defined boundaries.

For thousands of years after humans and beavers first met in North America, for First Nations (Canada) and Native American (United States) people, the beaver, through its incorporation into traditional origin stories and spirituality, and the social divisions that forged kinship ties, played an important role in forming Indigenous people's worldviews. By tracing lines of descent to a spirit animal with special powers, clan membership determined who your relatives were and to whom you could marry. For Indigenous people who recognize life spirits in natural things, stories and parables can encourage an interspecies connection that forges more porous boundaries between people and other species. That kind of worldview conflicted with the more rigid one understood by European settlers who saw humans as superior to and distinct from all other species, and whose spirituality was rooted in Christian monotheism.

Indigenous people not only consumed the flesh and fatty tails of beavers but relied on their tanned pelts for warm clothing. Since a diet lean in caribou or moose needed the help of beaver fat to improve absorption of protein, their food value was tied to the large number of calories beaver meat provided as well as its fattiness. For First Nations people, a toolkit that consisted of beaver-bone tools, beaver-tooth gambling dice, and a meat-carrying bag made from caribou and beaver, as well as winter garb made of beaver fur, suggests beavers and humans adapted to North America over a long period of time.[4] Not only were beavers critical for subsistence and warmth, but indigenous people also relied on the beaver's ability to naturally store water. Beavers create their own habitat by building dams and digging canals that slow water down to form ponds, deep pools, slow-moving side channels, and valuable wetlands, while the water spreads out to form meadows over time that create critical habitat for other species. Of particular importance to Pacific Northwest communities is how the wetland ecosystems beaver activities create are indispensable to fish nurseries that allow baby salmon to both conserve energy and evade predators like great blue herons. By impounding water, beavers also create a zone for surface water and shallow groundwater to mix. This zone, called the hyporheic zone, allows surface water to percolate through sediment and porous spaces, which reduces concentrations of pollutants. The percolating process also helps keep streams hydrated longer and to potentially flow at lower water temperatures, a real benefit to cool-water fish like salmon and trout. In the days when North America could claim millions of beavers, the Pacific Northwest was rich in beaver-built salmon habitat, with Washington's Stillaguamish river-basin habitat sustaining upward of seven million juvenile coho (salmon) each winter.[5]

With the arrival of Europeans, the beaver story that had typically been based on a blend of tribal traditions that included killing them for domestic consumption and trade goods to maintaining a sustainable subsistence-based human-beaver relationship[6] shifted overtime to the competitive killing of beaver for profit. By the 1400s, with the arrival of cod fishermen from France, Portugal, Spain, and England, an informal trade of furs and trinkets began between Indigenous North Americans and European fishermen. Native American and First Nations people can be credited with introducing the beaver with its warm and water-resistant fur to these fishermen, and from that simple exchange, the face of North America was changed forever. After the cod fishermen, others like the French explorer Jacques Cartier arrived, who, in the 1530s and 1540s, engaged in fur trading along the Saint Lawrence River. He looked at fur as trim and decorations for other products but failed to recognize the value of the actual pelts. The low-key informal nature of this early trade was short-lived, for when the European traders returned home, customers flocked to the docks to buy their beaver furs. From this, merchants quickly realized there was money to be made from New World furs for Old World hats. The fur trade was the economic foundation for European settlement in North America and by the 1600s was ready to swing (see figure 6.1).

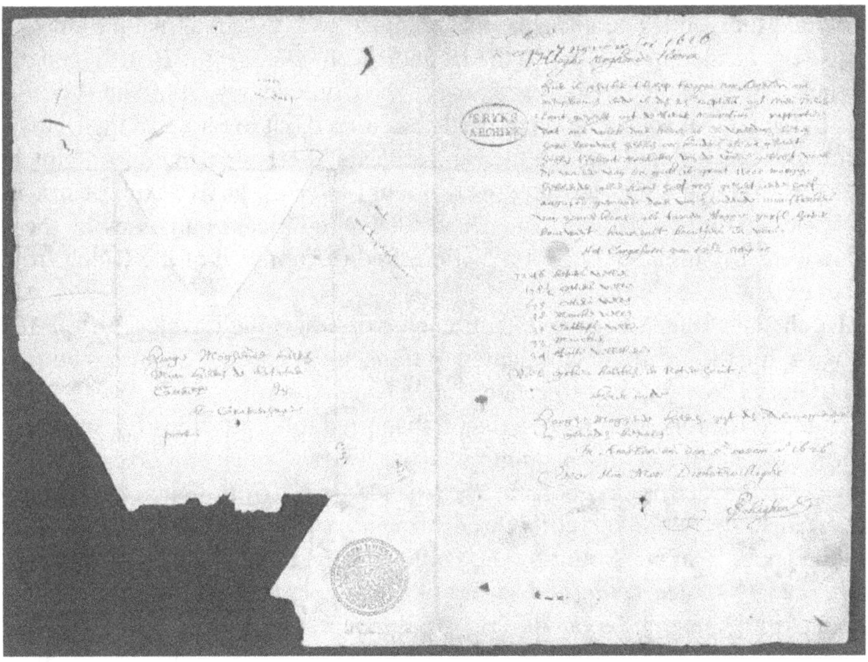

Figure 6.1 The Schagen Letter of November 1626 Regarding the Sale of "Manhattes Island" to the Dutch. Included in the letter was a list of cargo transported on the ship *The Arms of Amsterdam*. That cargo consisted of 7246 beaver belts, along with otter, mink, lynx, and muskrat pelts, as well as oak timbers and nutwood. The letter was on display at the New-York Historical Society Museum in an exhibit New York Before New York: The Castello Plan of New Amsterdam. May 2024. *Source*: Barbara K. Jones.

The demand for fur and eventually castoreum[7] was so great that a system rooted in money and markets spread inland, taking with it more and more First Nations people and the beaver they had relied on for centuries. Initially, First Nations people living near trading posts acted as either producers or middlemen who sought out beaver furs in more remote places; over time, though, the European fur traders moved past the established frontier to make more profits by dealing with producers directly. Through these activities, the boundary between traditional users and fur as a commodity rearranged itself often as beaver were eradicated and the center of trade moved westward and northward, and sometimes southward even though southern furs were of lower quality. As the demand for beavers accelerated, an intense rivalry between the French, English, and Dutch fur-trading operations in North America took shape. To capture more and more of the beaver market, the fur-trading operations put tremendous pressure on their Indigenous suppliers. This pressure revived past intertribal slights and introduce new ones, leading, in this atmosphere of economic rivalry and political upheaval, to the Beaver Wars. These wars were fought over land, beaver, and the trade that provided Indigenous people, like the Iroquois, with essential metal tools and guns. They ended in 1701 when the Iroquois Confederacy, the British, and the French agreed to the Great Peace, where the Confederacy would end hostilities in Ohio Country and allow those tribes forced out to return to their lands. Although this officially ended the Beaver Wars, there were many subsequent skirmishes as land and territory were fought over by Native Americans, as well as the French, British, and Dutch. Two significant outcomes of these wars were a realignment of First Nations and the removal of the Dutch from what is now called New York.

By the late 1600s, voyageurs, a French word meaning traveler, were hired to work for the fur trade companies to transport trade goods throughout the vast Great Lakes interior to rendezvous posts. At the posts, the goods were exchanged for furs, which were then transported to larger cities for shipment to the east coast. At the beginning of the season, when the lakes thawed, voyageurs carried trade goods as they headed west from Grand Portage on Lake Superior. While traveling through the area, they used *canot de nord*, or north canoes, that were built by the Ojibwe. The canoes were made of birchbark, cedar, spruce resin, and watap or stringy spruce roots; they weighed around three hundred pounds and were about twenty-five feet long and four-feet wide. The voyageurs operated in a self-sustaining system where their trade goods not only rewarded them with beaver pelts but with rice and smoked fish as well. Other hides and pelts of muskrat, deer, moose, and bear could be found mixed in the pelt bales that averaged ninety pounds each, with the men expected to portage at least two bales each time they walked the portage route. The beaver trade was obviously not for the faint of heart or weak of limb.

In 1670, England's King Charles II issued a royal charter to establish the Hudson's Bay Company (HBC). This company gave an exclusive trading monopoly over the entire Hudson Bay drainage basin to the governor and Company of Adventurers of England trading into Hudson Bay.[8] The King named his cousin Prince Rupert the first governor of the territory and in his cousin's honor called the drainage area "Rupert's Land." As a tribute to the King and his heirs, if they entered Rupert's Land, the company was obligated to provide them with two black beavers and two elk. Even though Rupert's Land covered nearly forty percent of what today is Canada, it took another two-hundred fifty-seven years before that tribute ever had to be paid when the Prince of Wales stopped by in 1927 enroute to his cattle ranch.

The tension between the British and the French led to a continuous jostling for territory and furs. Key to HBC's success was the strong trading relationships they built with First Nations, and later Inuit and Métis (people of mixed Indigenous and Euro-American ancestry) communities. The company relied on Indigenous expertise for general survival, which included surveying, trapping, translating, and eventually kinship when many European men married into First Nations families. After the Treaty of Paris in 1763, the French were out of the fur business, to be replaced by traders of Scottish and English descent, along with others like John Jacob Astor's American Fur Company, its subsidiary the Pacific Fur Company, and the Missouri Fur Company. However, through the 1870 Deed of Surrender, the company transferred from the United Kingdom most of its land rights, which also included Rupert's Land, to the newly formed country of Canada.

From the seventeenth century into the late nineteenth century, the European craze for beaver-felt hats kept beaver trappers in the business of killing beavers (see figure 6.2). The killing increased across the United States and Canada when metal traps and beaver castoreum as beaver bait were introduced by European traders into the lives of Native people. A combination of new technology and the absence of any rules or institutions that managed the number of beaver killed led to a near tragedy of the species. As a commodity, beaver furs not only made savvy Americans rich but, through the globalization of the fur trade, drove colonization. The beaver as a commodity metaphor made it not only possible for companies like HBC and individuals like John Jacob Astor to convert their "beaver product" into tremendous wealth, but it also led to a rapid decline in the number of "beaver districts" that had defined landscapes across the United States and Canada.

For all that resource extraction, the beaver was one of only a few animals that was also used as a form of currency. Before there was an official American currency, a deer hide or a whole deer, depending on its quality, served as

Figure 6.2 Beaver Trapping Display at the Cooke City Visitor Center. Cooke City, Montana. *Source*: Barbara K. Jones.

a form of currency. This led to the dollar eventually being called a buck. That same logic applied to the beaver, where in the 1800s, beaver, according to explorer and fur trader David Thompson, "were the gold coin of the country, with which the necessities of life were purchased."[9] The standardizing of beaver pelts was formalized by the HBC in 1670, when they introduced a North American unit of trade called "made beaver." The made beaver was equivalent to one prime-quality beaver pelt so that, until 1810 almost everything bought or sold at trading posts throughout the empire was priced in made beaver terms. In the 1860s, brass trade-tokens stood in for the actual beaver. As with any global market, this became a major problem for the West India Company's exports of beaver pelts to Europe. In New Netherlands (the first Dutch colony), where the company exported its furs, the beaver was not just a commodity but a negotiable instrument like our paper currency is today and the standard by which other significant money substitutes were measured. To devalue the beaver was to devalue the currency, yet for West India Company shareholders working in Amsterdam, an overabundance of furs had driven down their value. For a globalized market in beaver fur, the American form of exchange was not flexible enough to withstand the fluctuations in value that occurred outside the North American centers of trade, leading to a decline in made beaver as a unit of accounting.

In the American West, between the 1820s and 1840s, to extend the British monopoly on beaver trapping, the HBC went so far as to "scorch the streams" to destroy the beaver resources south of the Columbia River in Washington.[10] This policy created a "fur desert" boundary that kept separate American and British fur trappers. Through this scorched wetland policy, the HBC won the "western beaver wars," allowing them to acquire the thicker, more valuable pelts from British Columbia and Alberta without competition from American trappers. Their exclusive control over vast areas rich in fur-bearing animals led to the development of a trading enterprise that lasted almost two centuries.

Mountain men who from 1825 to 1840 operated independently as trappers put additional pressure on the beaver resource. They sold their furs at annual gatherings called rendezvous that continue today, with trappers and hunters holding annual meetings throughout parts of the Intermountain West to sell their furs, skulls, and antler sheds. Most recently, during Elkfest 2024 in the town of Jackson, Wyoming, the downtown streets were closed to accommodate the people and the thousands of antlers, skulls, and furs hunters, trappers, and the Boy Scouts had for sale.[11] Today, the Elkfest and the Boy Scout Antler Auction run in conjunction with the Mountain Man Rendezvous (see figure 6.3).

By the 1840s, trappers had practically exterminated the beaver in a large part of the American West, yet for the HBC the best was yet to come. In the

Figure 6.3 May 2024 Shed Sale during Elkfest in Jackson, Wyoming. *Source*: Barbara K. Jones.

two and a half years from 1854 to 1856, the HBC's catalog of sales indicated that 627,625 beaver skins were sold. The persistence and economic success of the beaver fur trade remind us of the critical role beaver trapping played in the colonization and settlement of both the British Provinces in Canada and the United States.[12] By the beginning of the twentieth century, the beaver, which was once, like the bison or passenger pigeon, a defining animal of North America was gone or so rare the species became difficult to find. When a population that once stood in the hundreds of millions fell to about 100,000, the feeling was that the beaver was on a path to extinction. Due to an increased awareness of other species disappearing, that fear was a legitimate one.

In what can be considered today the crossroads for the beaver's long-term survival, a few good beaver men stepped in. One was a British trapper living in Canada named Archibald Stansfield Belaney, who was more commonly known as Grey Owl, and the other was an employee of HBC at Rupert House named James Watt. Both had practical financial reasons to initially protect the beaver that over time changed to ones motivated by more lofty goals. Near the time Grey Owl was going to be excluded from trapping beaver and otter because he was not a "resident Indian"[13] as the government of Ontario required, he developed a conservation ethic that recognized the beaver's presence as vital to what he saw as wilderness. To shift people away from the "beaver as a commodity" understanding of beaver, he relied on messaging that spoke of "beaver as wilderness" and their absence as wasteland. By relying on such metaphors, Grey Owl helped move the wildness of beaver into the public's imagination and, through his writing, his relationship with "pet" beavers, and even his role as a "beaver wrangler,"[14] he offered people a new way of seeing these animals.

James Watt recognized that if selling beaver was how you earned money to survive, killing them all made no sense. He believed that if the beaver disappeared, the Cree people of northern Quebec would not only lose their traditional livelihood but also a critical aspect of their now interdependent economy. Since the Cree no longer depended only on what the land provided but were embedded in an economic system that relied on trapping and selling beaver to traders to survive, the absence of beaver would lead to their economic collapse, while for the HBC, fewer beaver meant a loss of both furs and customers. By relying on a strategy that had been done by HBC nearly one-hundred years earlier to significant success, in 1930, Watt bought two occupied beaver lodges from a trapper whose ancestral lands contained the lodges. With this purchase, he planned to leave the beavers alone to reproduce. To ensure a successful recovery, James and Maud Watt petitioned the government of Quebec to set aside land for a beaver preserve. Maud Watt convinced the deputy minister that a preserve managed together by the Watts working for the HBC and the Cree was necessary for the survival of both the Cree and the beaver. Their petition was so successful that they were given control of all the beavers in the Rupert House area, which consisted of over 7,200 square miles. After five years of no trapping, the Cree noticed a beaver population increase and they agreed to another five years of beaver recovery. Ten years after this effort began, controlled removals were allowed with 450 beavers taken. By 1944, the area sustained a population of over 13,000 beavers. The success of this management strategy inspired the formation of similar preserves and sanctuaries in the Northwest Territories and Ontario, extending the HBC-managed sanctuaries to more than 43,000 square miles. Seven years later, there were eleven connected beaver reserves in Quebec alone that consisted of more than 187,000 square miles. By the time of his death, James Watt's conservation efforts earned him the moniker "Beaver Chief" by the Cree people.[15]

By forming an institution that offered structure and rules, James Watt provided a community of beaver users with an approach to managing their common property resources sustainably. Today we see those strategies through Elinor Ostrom's work on "governing the commons," where she demonstrated how it is possible for communities to avoid a commons tragedy by collectively self-governing their "common pool" resources. In the 1920s and 1930s, where dominion over nature was still the defining story, such a strategy was considered an unusual or novel idea. No one did it for bison, passenger pigeons, elk, or even cod, but Watt and to some degree Belaney did it for beavers. By managing the beaver's value as an essential economic resource for many in beaver habitat, Watt relied on the economics of the beaver to protect them, while Belaney, through education and public outreach, made a case for the beaver's ecological role that allowed a new story to be told. Through the efforts of these two men, it became less likely the beaver would disappear forever from our shared lands.

ECOSYSTEM SERVICES: THE BEAVER BENEFIT

For natural capital to be an economic reality, a process that assigns wild nature a value in the marketplace is critical. Since nature and biodiversity have historically been treated as public goods, the rules of scarcity that typically encouraged us to change our behavior did not exist. In many areas, we extirpated sea otters, beavers, wolves, seals, peregrine falcons, and certain species of whales all because we could and because there were no social or economic consequences to that slaughter. Even as beavers began to recover from their role as commodities, by using pest descriptions and metaphors, we found new ways to marginalize them. That meant the real cost of killing most of the beavers wasn't realized until more recently, when, with a changing social attitude toward wildlife and a fragile recognition of our role in climate change, a sense of guilt, and the threats of flooding and drought became our new reality. As more and more people began losing their homes and businesses, as well as public infrastructure like roads and bridges to floodwaters, or crops began to wither and rivers dried up from drought, the economic costs associated with this increased flooding or poor water storage were being recognized and assessed through the marketplace. At one time, a solution to decrease flooding might have been to build dams, but they too have significant costs associated with their construction.[16] A more natural alternative would be to reintroduce beavers so they can build retention ponds that store and slow water or improve water quality to encourage wildlife to return, but that choice has different costs. For most of the twentieth century and on into the early twenty-first century, maintaining habitat for wild nature had a social and economic cost many people were not willing to bear. Yet today, as the connection between human well-being and healthy ecosystems is becoming clearer, the case for maintaining an ecosystem's services[17] is becoming a powerful way to manage our natural capital. With climate change a shared reality across diverse groups from farmers to fishermen to coastal residents, the potential for beavers to use their magic to mitigate this change has become a more likely one (figure 6.4).

Animals like beavers, bison, and wolves are examples of wild species that provide an array of ecosystem benefits, including services from each of the identified ecosystem service categories (regulating, supporting, provisioning,

Figure 6.4 A Beaver Dam in Grand Teton National Park. The dam stores water that has created important habitat for many species (from moose to cold water fish to waterfowl). May 2023. *Source*: Barbara K. Jones.

and cultural). Yet, in the time it took us to figure those roles out, it was almost too late. Beavers were overharvested in Canada, the American north, and the Pacific Northwest, leading to a decline in beaver ponds. These ponds not only provided habitat for fish and other wildlife, but they also reduced downstream flooding. Once the beavers were gone, humans replaced the services beavers provided for free with expensive concrete dams (see figure 6.4). In the Columbia watershed in Washington state, proposals for more dams were brought forward, but the Washington Department of Ecology challenged the need for dams by releasing, in 2013, beaver back into the Methow Valley (where beavers had been eradicated). Today there are 176 beaver ponds at fifty-one sites storing millions of gallons of water. According to the Methow Beaver Project report, the beavers' return will provide a measurable, lasting benefit to the watershed.[18] Just as returning bison to small pockets of the plains has taken decades, it has also taken us a long time to bring beavers back to sustainable levels. Over a century after bison numbers crashed, small, sustainable herds of bison have returned to once again spread their grazing and manure over large areas. Their sharp hooves are cutting into the soil again to aerate it and create "tiny buffalo ponds"[19] for migrating birds, while the wolf's return to YNP has helped ecosystem's recover by not only reducing overgrazing from excessive numbers of prey species butcreating nutrient-rich "hot spots" through the presence of carcasses from their kills. The critical natural services these species provide were almost lost because we were slow to recognize just how incompatible our self-interests have been with the needs of the natural world.

Returning beavers or bison to the land is what some would call "environmental pragmatism," where solutions to resource problems can be generated by working with human and wildlife needs in a complementary way. Tied to this pragmatism is the ecosystem services (ES) approach that tends to offer stakeholders a voice in the process of preservation and conservation. The inclusive nature of ES leads to, in many cases, more positive and productive outcomes for those species affected. To the "environmental purists," the ecosystem services model tends to rely too much on satisfying the human in the room and not enough on keeping spaces free of our influence or in maximizing biodiversity. Taken together, though, both approaches have tremendous value. Pragmatic solutions have broader appeal because they can encourage decision-makers to consider nature-based strategies if they see the economic value in those choices, while purist solutions remind us of a species' intrinsic value. In 2012, after Superstorm Sandy hit the east coast of the United States, the resulting severe flooding and property loss made stabilizing and reinvigorating coastal sand dunes a critical strategy for many coastal policymakers. One pragmatic solution to dune stabilization was to rely on repurposed Christmas trees, while another was to use nature-based solutions, like oyster reef breakwater structures.[20] These structures not only diminish wave energy, but they also filter water and provide habitat for key commercial and recreational fisheries.[21] Relatively simple, natural, and cost-effective methods like these can help address the problems coastal residents will continue to face from climate change.[22]

Unlike human-engineered resilience akin to dam building or rechanneling rivers, the beaver's engineering skills help enhance an ecosystem's natural resilience. Natural resilience is superior to engineered resilience in many ways because, unlike natural resilience, which is proactive, engineered resilience is reactive. Proactive actions maintain or build on natural assets and functions, while reactive actions replicate a function or change our expectations in response to a loss of an ecosystem service benefit. Natural resilience is also more comprehensive and systems-based, while engineered resilience reduces the system to its parts and offers discrete benefits based on replacing a missing piece without understanding its role in the entire system's overall health. A final and extremely important aspect of natural resilience is its cost. Natural resilience efforts tend to be less expensive in the long run because the effort expended is done by a self-regulating ecological system rather than through the efforts of human labor and built capital.[23] That difference means the ecosystem services beavers provide, including water storage, fire reduction, channel protection, improved riparian corridors, salmon restoration, and flood reduction, as well as the cultural services of tourism, fishing, hunting waterfowl, and trapping, are more durable and cheaper to maintain than if we tried to engineer ourselves into healthier and more resilient wetland and riparian ecosystems.

Salmon benefiting from beaver activity has long been contested, yet recent studies have demonstrated that beaver dams create essential nurseries for salmon. For ranchers and farmers, climate change effects that include drought make keeping water on the land longer critical, just as improving riparian corridors by spreading water out improves ranchland habitat. So livestock have access to more water, ranchers are being encouraged to partner with state and federal fish and wildlife biologists to "re-beaver" their creeks.[24] For outdoor sports enthusiasts, expanding wetlands and increasing stream flows and sedimentation provide better fishing and wildfowl hunting opportunities, while wildlife viewers and recreationists get to experience more resilient and intact ecosystems. For our mutual well-being, beaver ponds press water into the earth, a process that not only helps filter out heavy metals and pollutants but allows water to flow into subterranean channels that, as it cools, seeps to the surface. That seepage can increase summer water flow and lower stream temperatures.

Dam building also creates wooded wetlands. Runoff from snowmelt can be captured, stored, and then slowly released from beaver wetlands during the summer when water levels are at their lowest. Moreover, these wetland areas beavers create can reduce fire risk, allowing them to join the famous fire prevention bear as his sidekick, "Smokey the Beaver." According to Torrey Ritter, Montana's FWP regional nongame biologist, "Beaver wetlands also act as a type of Noah's Ark, where small mammals, frogs, birds, and other animals can escape fire."[25] A recent Facebook post by the Western Watersheds Project included a picture of a raging fire with elk seeking refuge in a river. The headline of the post read, "With their home a raging inferno, these elk would have died without the river. Beavers expand the size of a river safe zone by creating wetlands and ponds."

In the United States, beaver tourism is not yet on the radar of many ecotour providers, but in Great Britain, where beavers have only recently been reintroduced after an over four-hundred-year absence, beaver tourism has become a new revenue-generating opportunity. Beaver tourism is also growing across Europe from Denmark to Poland, where beavers play an important role in wildlife tourism receipts. For over twenty years organized beaver tours into the Klosterheden Forest in Denmark have taken people to see beavers. In Belgium, through a marketing campaign based on the themes of beer, beavers, and castles, beaver watching holidays like a beaver walking trek through the forests of the Ardennes that ends at La Chouffe craft brewery are becoming popular. In many places, beavers have been reintroduced to encourage tourism into areas where industry and agriculture are no longer key revenue generators. In the Breton Village of Brennilis, guided tours to the town's beaver museum to see dams, lodges, and wetland habitats are offered, while in the Netherlands' Biesboch National Park the first itinerary choice on their website is to go sailing, cycling, or walking to spot beavers. The park sells beaver-related merchandise in its gift shop and offers boat tours at dusk to look for them. In Naliboki Forest in Belarus, there are approximately four-thousand beaver ponds and about three-thousand marsh plots at the sites of former beaver ponds. Seven-to-ten-day beaver tours are offered where guests travel by foot, kayak, and car to see the beavers. Although beavers are a nocturnal species, they tend to be most active at sunrise and sunset, making them crepuscular as well. Because of this, a visit to see beavers can require guests to stay overnight. An overnight stay is helpful to local economies because it supports local lodges, restaurants, and shops.[26]

Beaver enthusiasts, along with the government of the United Kingdom, have been motivated to reintroduce beavers because of their ecosystem engineering skills. Those skills have led to biodiversity increases, the slowing of water, water quality improvement, and ecotourism opportunities. Today, a group called Perthshire Wildlife offers Winter Beaver Safaris or Riverbank Beaver Safaris along the River Tay in Scotland, while in March 2021, the first beaver pair was reintroduced to the Cors Dyfi Nature Reserve in Powys,

Wales. Alicia Leow-Dyke of the Welsh Beaver Project hopes the mammals will bring a "tourism boost" to mid-Wales. According to Leow-Dyke, "There is a lot of interest around beavers and tourism, and we've already had people asking when they can come and see them. We have seen it with other projects in Scotland and England, where there was huge tourism interest that coincided with that work." [27]

A recent promotion for a beaver ecology tour run by the Pinelands Alliance in Shamong, New Jersey, suggests beavers are beginning to move onto people's radar in the United States as well. This tour takes kayakers to Goshen Pond to see how these "furry engineers"[28] have reclaimed their status as a keystone species in pineland ecosystems. The tour covers both natural and cultural history in that the guides speak to the beaver's role in maintaining the Mullica River habitat as well as their extirpation and the successful efforts beginning in the 1930s to bring them back. In the online publication *Untapped New York,* a spring 2021 headline reads: THE BOOGIE DOWN BEAVERS OF NYC'S BRONX RIVER. The surprising part of this headline is that, even

Figure 6.5 This Is John B. Pine's Example of the New York City Seal with Its Two Beavers. This seal is included in his 1915 book *Seal and Flag of the City of New York,* which in its entirety is in the public domain. A similar version is on the front of P.S. 334 in New York City. *Source*: John B. Pine.

though New York City's iconography is loaded with beavers, from New York City's Seal to Beaver Street to City College's mascot of Bennie the Beaver, for two hundred years there have been no beavers in the city. They had first been hunted to extinction and then, when the Bronx River became an open sewer, any remaining wildlife moved on. After years of remediation work by the Bronx River Alliance, in 2007, someone spotted a beaver on the river. He was named José, in honor of U.S. Representative José Serrano of New York, a champion of the Bronx River's remediation. Then, in 2010, a second beaver was spotted. With this one, a naming vote was taken and by an overwhelming number, the name "Justin" in honor of Canadian singer Justin Bieber was selected. By 2017, a third was spotted and now, around sunset, you can see beavers paddling around the Bronx River. Typical for New York City wildlife, José the BX Beaver even has an X/Twitter account with 172 followers. The return of river wildlife like the beaver has encouraged the Bronx River Alliance to offer a guided canoe tour, reminding us that ecotourism is another way to measure the value of the ecosystem services beavers provide to both nature and our well-being(see figure 6.5).[29]

BEAVER MANAGEMENT FOR COEXISTENCE

In an article called "Beaver Restoration Across Boundaries," Rachelle Haddock of the Miistakis Institute in Calgary, Alberta, wrote about the role of transboundary partnerships in beaver restoration. Although these transboundary partnerships sound a lot like coexistence across groups of competing users or even species,[30] the language or even the expectation of coexistence is typically absent in many beaver mitigation efforts when lethal or destructive strategies are easier and more readily accepted. Just as naming the American bison America's national mammal in 2016 did little for the wild status of bison around the United States, naming the beaver Canada's national mammal in 1975 didn't do much for the beaver in Canada either. Lethal efforts at eradication with guns and traps remain the norm, with only about one-quarter of Alberta's municipalities using nonlethal means to handle beaver conflicts.[31] As part of a beaver restoration project in Canada, researchers sought out examples of beaver coexistence, translocation, or relocation cases in the western United States. They chose six different sites that offered different policy approaches from not coexisting at all by relying on long-standing trapping regulations to the highest level of coexistence that resulted in implementing a state beaver bill. The six states the Canadian group selected were: California, Oregon, Wyoming, Utah, New Mexico, and Washington.

For California, with the lowest level of coexistence efforts, the group interviewed Kate Lundquist of the Occidental Arts & Ecology Center WATER

Institute. There, the beaver story ends like so many others with their near disappearance by the beginning of the twentieth century. In California's Central Valley, over ninety percent of the wetlands disappeared to become farmland, while market hunting took an additional toll. Over concerns from sportsmen regarding lost hunting opportunities, the California State Fish Commission was renamed the Division of Fish and Game. This change reminds us of the lens through which wildlife is conserved, and to Steven Fountain, professor of history at Washington State University, that lens is through conservation of game, not restoration of wildlife or entire ecosystems. The Great Depression saw increased efforts at game and fish restoration, with beaver restoration being an unusual effort because they had significantly fewer supporters than fish and game did. When beavers were restored to the mountains, sportsmen and farmers saw this as returning the animal to its "rightful place," but when the same animal appeared in the levee-strewn Delta or the canals of the agriculturally rich Imperial Valley,[32] they were, once again, vermin.

Conflicts like this one led the U.S. Forest Service to include an official division of "good" and "bad" animals in its newsletter for the California region. This division supported the Department of Agriculture's Biological Survey's (later Wildlife Services) efforts to eradicate varmints. Fur bearers, with no mention of beavers specifically because they were uncommon by then, were popular, as were other animals that generated profit. An idea to make beavers profitable again rested on them being "wholly controlled if not [in] a fully domesticated condition." The unlikeliness of domestication and private ownership of profit-generating beavers meant that, like carnivores that destroyed game or animals that destroyed crops or timber, beavers entered an expanding category of "bad" species called varmints. Although deer and waterfowl did more damage to agricultural crops and timber than many of the "bad" animals like beavers or porcupines did, the feeling that game animals more than compensated for agricultural losses encouraged farmers to tolerate them.[33] In 1911, thinking beavers were nearly extinct in California, a California law was passed to protect the beaver. Within six years, as the population recovered, conflicts with the agricultural community over these "pests" led to a "de facto open season" on beaver.[34] In support of property owners, laws were "relaxed" so that beavers could still be killed. That law remains in effect today. By the mid-1920s, after recognizing the benefits of having beavers around, the California Department of Fish and Wildlife started a relocation program. For twenty-seven years, beavers were relocated around the state, with many of today's beavers tied to those earlier relocations.

In 1937, the preeminent California naturalist and zoologist Joseph Grinnell published his definitive tome on fur bearers in the state called *Fur Bearing Mammals of California.* That book included the historic ranges of California's native fauna and has remained the source for defining what species are native to the state. In 1942, Grinnell's contemporary, naturalist Donald Tappe, added to that canon of knowledge by concluding that beaver had a limited range in California and never inhabited much of the state's coastal areas, including the San Francisco Bay Area or the Sierra Nevada mountains.[35] Grinnell and Tappe's evidence was based on the *absence* of beaver and can be understood as an example of what fisheries biologist Daniel Pauly called "shifting baseline syndrome." The acceptance of degraded ecosystems as the standard occurs when scientists use depleted, absent, or over-abundant population numbers for a species that existed at the start of their careers as the baseline for the ecosystem rather than using population numbers for a species in its natural state. According to Aaron Hall of Defenders of Wildlife, the environmental baseline for beaver shifted from eighteenth and nineteenth-century America when "beaver would have been everywhere so the landscape settlers saw would have already been modified by their loss."

This reality allowed Grinnell and Tappe to draw historic beaver distribution maps based on their assessment of habitats they saw as suitable for beavers. Fast-forward to today, and beavers are still currently managed as a non-native pest in much of the state. As climate change and water shortages became a more critical aspect of life in California, a group of beaver advocates, wildlife biologists, and naturalists came together to change the beaver's enduring legacy as a non-native pest to one of a native ecosystem engineer.[36] The group, which included Kate Lundquist, relied on specimens in museums and archaeological sites, historical fur-trapping records, and newspaper accounts, as well as geographic place names and Native American tribal names for "beaver" to demonstrate that beavers were native to the Sierra Nevada and most of the coast, all the way down to San Diego. This evidence is helping change the minds of people throughout the state, but not the regulations regarding their management. When the state Wildlife Protection Act of 2019 was passed, recreational trapping beavers in California became illegal. That rule did not apply to landowners or tenants, who can still apply for a permit to kill beavers perceived as causing damage. The permit holder must document the damage but doesn't have to report how many beavers are killed.

According to a report by the Water Institute in California, even though beavers provide a myriad of ecosystem services to the state's watersheds that a coauthor of the report, Broc Dolman summarized as "slow it, spread it, sink it, store it, [and] share it,"[37] there are currently no California Department of Fish and Wildlife (CDFW) codes that promote beaver stewardship and/or restoration. Most rodents, including beavers, have been designated as "detrimental species" because the Fish and Game Commission (FGC) and the California Department of Food and Agriculture have determined that their importation, transportation, or possession could pose a threat to native wildlife, the agricultural interests of the state, or public health and safety. For this reason, under FGC section 2150, special permits cannot be issued for the possession, import, or transport of beavers. According to Lundquist, this law leaves a lot of room for interpretation. Even though the CDFW webpage now recommends "nature-based solutions" for managing beavers and there are links to *Living with Beavers*, *Beaver in California: Creating a Culture of Stewardship*, *Occidental Arts and Ecology Center Water Institute*, and the *Sierra Wildlife Coalition*, the absence of clearer codes regarding beaver management makes Lundquist wonder to "what extent the Department is informed about the potential benefits of having beaver on the landscape and is committed to managing them for those benefits."[38]

In contrast to California's approach to inconsistently pursuing human-beaver coexistence are those in Utah and Wyoming. Wyoming recognizes the value of beaver through the lens of the many state benefits they provide, which include trumpeter swan restoration, fisheries for maintaining year-round stream flow, improvement of degraded riparian corridors for big game, and keeping water on the land for private landowners. For the Wyoming Wetland Society (WWS), improving wetland habitat for the trumpeter swan, which in Wyoming is designated as a species of greatest concern, is their mission. This mission works well for beavers because they are wetland engineers. In other states, beaver coexistence efforts are also tied to potential endangered species' needs, such as those with the Columbia spotted frog and western toad.[39] Conversations with ranchers and landowners in Wyoming suggest that the repercussions from listing a species are a significant worry because an ESA listing or the designation of critical habitat for a listed species can result in complicated land-use restrictions. The Wyoming Safe Harbor Program can protect landowners from this liability if they create or improve habitat for listed species, which for ranchers, coexistence with beaver has the potential to proactively mitigate future private or public land-use restrictions.[40] In Wyoming, beaver benefits are limited to the functions they provide and, according to the Wyoming Wetland Society webpage: "The awareness is there that beaver are important from a keystone species perspective. Beavers are a very useful and inexpensive tool for creating wetlands. Even if the beavers are killed out, the dams they built continue to trap silt, maintain stream flows and temperature regimes that benefit fish species."[41] For the Wyoming Wetland Society, their priority is the trumpeter swan and the services beaver provide, but not necessarily the presence of beaver.

In Utah, wildlife management plans are usually for big game or for protecting species that have the potential to be listed on the endangered species list. The state added a third reason with a beaver management plan that was the result of a landowner, a USFS National Forest supervisor, and some local citizens urging that beaver be returned to Forest Service lands. The feeling from this group was the nation's forests could benefit from reducing flood damage, reconnecting streambeds, and maintaining riparian habitat, which were all services beavers provided for free. Yet without shared rules across federal agencies, the efforts of one agency intending to return beavers to the forests could be undermined by those of another that allows beaver trapping.

For Oregon, where beavers are treated as pests, a long cultural heritage of trapping has clouded efforts by lawmakers to find a reasonable balance between two competing value systems that revolve around trapping as an individual's cultural heritage versus beaver as ecosystem engineers. To address this conflict, a conversation with Dr. Fouty provided me with the context of the "2020 emergency climate change and biodiversity losses response effort to close beaver trapping and hunting as a recreational activity on federal land in Oregon." On May 21, 2020, a group of concerned individuals, scientists (including Dr. Fouty), and conservationists submitted a request to the Oregon Fish and Wildlife Commission (OFWC) to close federally managed public lands and associated waters to beaver trapping and hunting in Oregon. That request was intended for consideration at the OFWC's June 12th meeting on furbearer regulations that is open for review and change on even-numbered years. The Commission received hundreds of letters both in support and in opposition to the requested change, with some of the letters in support coming from scientists in academic settings, the Wetland Institute, BARK, Audubon Society, and the Columbia River Inter-Tribal Fish Commission. In their letters to the Commission, those individuals relied on scientific evidence from the Oregon Department of Fish and Wildlife, USFWS Technical Reports, MS theses from Oregon State University and Humboldt State University, and peer-reviewed articles in *Ecology*, the *International Journal of Climatology*, and the *North American Journal of Fisheries and Management* to demonstrate how the many ecosystem services beavers provide can help mitigate the effects of drought and wildfire exacerbated by climate change. Some of those who opposed the change were the Oregon Dairy Farmer's Association, Oregon Forest and Industries Council, Fur Takers of America, Oregon Outdoor Council, Oregon Hunter's Association, and the Oregon Sporting Dog Association. Their letters curiously relied on the "lack" of scientific evidence that existed to support the change, with comments like "zero scientific evidence," "no scientific evidence supports trapping bans," or "[t]he notion that anadromous fish recovery is highly correlated with increased beaver population is flawed. At best, it is a socially constructed idea with little basis in science." Three additional comments were interesting because they added cultural heritage arguments to their responses: "There is no demonstrable science to prove [beaver will create more dams, 'ergo more Coho'], what we do know is beaver thrive in early seral riparian conditions, [while] most if not all modern forestry regulations and management plans intend to create exactly the opposite of those conditions in near stream areas." Another comment included a request to "not encumber an important wildlife management tool with such a blatant effort to support one value system at the expense of all others," while another wrote, "please stop the damage these [supporters of the change] are trying to make to our Oregon traditions."

By July 5, 2020, Oregon's wildfire season began and became the most destructive wildfire season on record in Oregon. Fouty, working with Ernie Niemi, President of Natural Resource Economics, wrote an economic analysis on the ecosystem services and the economic benefits of beaver-created and maintained habitat in Oregon. Their paper was brought before the Oregon Fish and Wildlife Commission on September 24, 2020, as part of the "Petition to Initiate Rulemaking to Amend OAR 635-050-0070 to Permanently Close Commercial and Recreational Beaver Trapping and Hunting on Federally Managed Public Lands and the Waters that Flows Through These Lands." According to Niemi and Fouty's report, beaver trapping and castor sales generated between the years 2015 and 2019 less than $44,808 and supported fewer than 170 individuals, while the ecosystem services value of the beavers' presence on those lands was estimated to be in the hundreds of millions to billions of dollars.[42] However, less than two months later, on November 13, the Commission denied their petition request in a 3–3 vote despite the many economic and ecological benefits that would result from ending beaver trapping on public land. For Fouty, Oregon's ongoing drought and the associated wildfires should have meant the "stars had aligned" and the petition would pass, but it didn't.

To the Commission, the beaver was not only not at risk of extinction or a species of greatest conservation need under the Oregon Conservation Strategy, but the existing rule was not unreasonably complex. In addition, no other state or the federal government has the authority to regulate furbearer harvest in Oregon; technological, economic, or other factors had not changed in the area affected by the rule; and the legal basis for the current rule regarding beaver trapping had not changed so the most recently adopted rule in June 2020 was still relevant. According to Derek Broman, the Oregon Department of Fish and Wildlife (ODFW) Carnivore-Furbearer Coordinator, the push to ban trapping on federal lands was not necessarily about protecting river ecosystems or rehabilitating salmon habit, but rather a "red herring" to distract from the petitioners' goal to end trapping altogether. Broman argued against the petitioners' request with the science that showed how little impact a small number of trappers had on the beaver; instead, he said, a lack of habitat and the presence of robust apex predators were the greater limiting factors.[43] Those in favor of the rule change used the beaver's role in mitigating climate change, carbon capture, wetland restoration, and increasing biological diversity to encourage an end to beaver trapping on federally managed land in Oregon. To them, the Commission's argument wasn't rooted in contemporary science and relied instead on the value of a trapping heritage to win. Bark, a nonprofit organized to protect Mount Hood National Forest and is a supporter of returning beaver to Mount Hood, used their organization's online newsletter to write how "Our state animal, university mascot, and ecological engineer extraordinaire is forced to sit on the sidelines during these challenging times, since it is valued by some state officials, and less than 170 people in the state, as merely a species to trap and shoot for recreation."[44]

After the vote, rather than admit beaver contributions were economically or ecologically significant, the Commission organized a Beaver Management Working Group (BMWG) with participants from various stakeholder groups that ran from June 2021 to April 2022. During the ten-month meeting period, rather than address the big topics of trapping and hunting closure, benefits to salmon, water, wetland habitat, and climate change, the working group chose to avoid them. By May 2021, an ongoing drought from little rain and low snowpack levels led to one of the earliest starts to fire season in over forty years. In June 2022, with little progress having been made from BMWG, a second request was made to end trapping and hunting of beaver on public lands; however, the request was ignored by the Commission. Most recently, in June 2023, ODFW released its 3-Year Beaver Action Plan with ten beaver emphasis areas selected. Of those ten, three were federally managed and seven were largely private. One of the three federal areas has historically been closed to trapping and hunting, while the other two remain open.

In Oregon and elsewhere, heritage is a powerful argument that relies heavily on metaphors and stories to maintain its persuasive power over that of coexistence. While the failure to end trapping on federal lands in Oregon is an example of the inconsistencies that exist between federal land managing agencies, it did offer both sides the opportunity for public engagement and education on two competing worldviews. Moreover, when Broman called the petition a "red herring" to end trapping, he was reminding the many non-trappers in Oregon of the power the "trapping as cultural heritage" metaphor has over state wildlife management decisions.

BEAVER "REWILDING"

Like the word coexistence, rewilding has linguistic baggage that leads us to ask the following questions: Are there distinctions between reintroducing and rewilding? And are they both susceptible to similar social limits? Human-human conflicts remind us that culture is contested and that while some people might love beavers, bison, or wolves living in their shared space, others might find the sheer size of a bison overwhelming or the tree-gnawing tendencies of a beaver annoying. If, though, we accept that people value nature differently and recognize that human-wildlife conflicts can generate human-human conflicts, by expanding coexistence's reach and encouraging a greater social tolerance for wild nature, it becomes possible to minimize the source of these conflicts and find common ground (see figure 6.6).

According to Dr. Roger Auster, a lecturer in Environmental Social Science at the University of Exeter, in his work on the human dimensions of reintroducing the Eurasian beaver in England,[45] a better way to see reintroductions as a form of rewilding is through something he calls "renewed coexistence." The language of "renewed" builds on what he sees as the established understanding of coexistence while recognizing the uniqueness of the reintroduction context. In his research at the River Otter Beaver Trial site in Devon, England, he noticed that for some people there is a "link" between the reintroduced beavers and "the people [who] put them there." The existence of this link ties the people responsible for the species' return to any conflicts resulting from that reintroduction. The perception of this human tie to the species' presence simultaneously diminishes its wildness just as it increases an expectation of human management responses. To Auster, the term "renewed" untangles that tie by highlighting a relationship that historically existed between humans and the reintroduced species. For beaver, by bringing them back to the land, an opportunity to renew the relationship between humans and the reintroduced species presents itself. Auster sees that relationship as key to encouraging human-beaver coexistence.[46]

Figure 6.6 Sterling, Massachusetts Police Found the Beaver Tail Nailed to a Tree below a "No Gnawing" Sign at Stuart Pond on April 27, 2023. The sign was posted to make people aware there were beavers in the pond. The tail was added later. *Source*: Town of Sterling Animal Control Facebook Photo.

The implication of renewing is for people to reconnect with a species they have had some experience living with before. Auster recognizes that, like with the beaver, the experience doesn't necessarily have to be a positive one, but sees renewed coexistence as an approach that, by addressing the human dimension from the start of the reintroduction, makes coexistence's persistence long-term more likely. For the North American beaver, the negative language that has already constrained it to live outside our invented boundaries is a challenge this approach would have to overcome because, by their very nature, human-created borders do not make particularly good coexistence bedfellows.

According to Aaron Hall, the language of nuisance and pest has made the social carrying capacity of beaver so low that the only direction social tolerance for beaver can go is up. Hall's comment suggests there is plenty of room to build social tolerance by changing the beavers' story. Since beavers are not predators there is less fear associated with their presence, and they have a charismatic face that makes people feel "warm and fuzzy." These traits, along with their role as a keystone species that can help mitigate climate change, have in many places already begun to shift the beavers' story. Yet beaver critics have taken on a disproportionate role in maintaining them as a nuisance, such that in Colorado beavers can be killed under a nuisance permit. By writing the Nuisance Rodent Act, which was passed to address a prairie dog problem, as a rodent law, beavers as rodents can be killed when they weren't even "on the radar." By nature of being rodents, prairie dogs and by association beavers have had a bad reputation in Colorado ranching communities. According to Mike Callahan, founder of the Beaver Institute, the rodent classification also allows people to lump beavers with non-native rats, further damaging the beaver's brand. Calling beavers "river rats" only serves to fuel the beaver-as-nuisance metaphor.

In 2021, Wildlife Services (APHIS) under the auspices of the Department of Agriculture killed 24,683 beavers across forty-two states, with an additional 125 freed, released, or relocated.[47] These killings were done to minimize human-wildlife conflicts that included gnawing on trees and building dams. Apparently, the beaver that floods the road, cuts down a tree, or builds a wetland in a rancher's field must be removed because it *is* a nuisance. Why coexist with a nuisance when there are relatively easy lethal means readily available to end the human-wildlife conflict? Coexistence is harder and requires a new way of thinking about "conflict species." This new way of thinking sees human-wildlife borders as porous, recognizes the ecosystem services the species contributes to our well-being, assigns intrinsic value to the species right to exist, and increases social tolerance for species recovery through education associated with efforts at reintroductions or rewilding.

Interesting new research done by Britt van Zelst addresses the challenges in balancing the beaver's ecosystem services with what she describes as its disservices. Brett and I met at a Beaver Conference in Maryland during the summer of 2022, where we both were presenting on topics related to the services and, for her, the disservices as well of beaver in a twenty-first-century human-defined landscape. Her subsequent graduate school research project that resulted from this research was called "Beavers in Anthropogenic Landscapes: a Blessing or a Burden?" Her case study was the Netherlands where beavers were reintroduced in 1988. In the Netherlands, since both the Dutch and the beaver are ecosystem engineers, there is an increasing need to find ways to sustainably manage potentially competitive human-beaver relationships, so that, as she describes it, "people can enjoy the [beaver's] blessings and simultaneously handle [its] burdens." For van Zelst, achieving a sustainable human-beaver coexistence would require not only overcoming a lack of knowledge about beaver effects specific to anthropogenic landscapes in the Netherlands, but developing an inter- and transdisciplinary study approach that included the involvement of both natural and social scientists, as well as other stakeholders. Of particular interest to me as an anthropologist is how she assigns value equally to the specific effects of beavers on both nature and society. She considers how on Dutch landscapes, addressing the social conflicts between beaver and the Dutch people, as well as the natural in terms of biodiversity, water quality, and water quantity, make management decisions particularly challenging.[48] By recognizing there are both blessings and disservices or burdens people must bear when living with beaver, van Zelst demonstrates a potential path to bridging this coexistence divide. This awareness becomes particularly important when we add anthropogenic landscapes to the conversation because human-modified landscapes might require us to rethink how the burdens and blessings associated with beaver in our shared spaces have changed to match our twenty-first-century ecological, economic, and social needs. Maybe water storage and wetland construction from dam building are not necessarily the burdens they once were, just as the blessing of beaver as furbearers with profitable pelts might no longer resonate outside a small number of trappers. Anthropogenic landscapes require us to rethink a lot of our assumptions about what today's social, ecological, and economic needs are.

For those working to rebrand beaver, their efforts in education, relocation, and beaver damage mitigation have been gaining precedence around the globe. Citizen scientists are forming beaver advocacy groups and using iBeaver, a data-collecting app that identifies existing beaver habitats, to generate important information about the species. To both educate people about beavers and to defend the Martinez (California) beavers, for the past fourteen years, Heidi Perryman, with her nonprofit organization called "Worth A Dam," has organized an annual beaver festival that now occurs in five other states and

Canada. To her, drought makes it "more important than ever to coexist with these important 'water savers.'"[49] Groups like Defenders of Wildlife and The Nature Conservancy, as well as local agencies, watershed groups, and beaver fans, are joining together to help form new beaver-inspired communities. Coexistence efforts are also becoming more popular as new technology and new understandings of beaver behavior become better understood. Ranchland habitat that has been too degraded or "too etched" is not good habitat for beaver, yet the beaver's skills are needed, so beaver-dam analogs (BDA) are built. These BDAs are man-made structures that mimic the form and function of beaver dams. To work, they may include composting across a streambed with sticks to get the water to back up and spread out, hoping beaver who, according to Aaron Hall, are "lazy-ass smart," take advantage of existing structures. Beaver flow devices, like beaver deceivers and pond levelers, are pipes that can be installed through beaver dams, or pipe and/or fence systems that protect road culverts from being blocked. With a flow device, beavers do not detect water flowing, so they are not inspired to capture it. By not capturing more water, the device reduces the chances people will be inconvenienced by road, culvert, or farmland flooding.

Mike Callahan started the Pioneer Valley Wetland Volunteers as a volunteer organization in 1998 to help people with beaver problems. In 2000, he formed a business called Beaver Solutions LLC and then seventeen years later, he founded the Beaver Institute, Inc. The Beaver Institute's work on outreach and education helps people learn to live with beavers. In Massachusetts, where the Beaver Institute is located, an anti-trapping law sponsored by animal and wildlife advocates passed with two-thirds of the vote. That change pushed the Beaver Institute to develop alternative nonlethal solutions to beaver engineering that occur in the "wrong" places. Relocating problem beavers has also been another important beaver mitigation strategy, while for trees, fencing or painting trunks with a sand-paint mix can minimize damage to a homeowner's or farmer's vegetation. These efforts, combined with the benefits beavers provide to us and our shared ecosystems, are helping to make beaver coexistence a more appealing option.

NEXT STEPS

An interesting approach to shifting people's thinking about pests, boundaries, and beavers is to consider our relationship with them through the lens of behavioral economic theory. Behavioral economic theory is not necessarily about money, but rather about decision-making whether the decision-maker is a stockbroker or a rancher. Classic economic theory assumes that humans are rational decision-makers, while behavioral economists don't see that as always true. Instead, people's biases can lead them to make decisions that aren't ideal for their own or society's needs. That positionality or bias can explain why people respond to specific wildlife in what appears to be irrational ways. A bias that marginalizes beavers might be explained through cultural heritage, family lore, past experiences, group pressure, or stories and metaphors that have relied on soundbites and catchy story characters to maintain a negative way of seeing beavers on the land. The fact that ranchers have significantly fewer problems coexisting with elk than they do with less costly bears or wolves is an example of how preexisting biases influence their behavioral responses. In Montana, ranchers pay to irrigate their fields for hay and cattle grazing but end up feeding extremely large herds of wild elk. That same tolerance is not shown to prairie dogs or bison, but since elk are tolerated, that behavior does not seem irrational (see figure 6.7).

In some New Jersey communities, beavers have returned. Although beavers gnaw on and chew down riparian trees and bushes and build small dams, their presence does slow water and can reduce flooding. Yet for these mostly positive contributions, beaver dams and lodges can be destroyed, and the beaver either killed or relocated. An extremely large beaver lodge in New Jersey that I have watched evolve in size as part of a healthy wetland ecosystem was recently destroyed. By destroying the lodge and then the beaver dam, the once lush wetland has turned into a dry, weedy lot. The costly presence of an extremely high number of white-tailed deer on New Jersey roads and in people's lawns and gardens rarely results in the same outcomes.

Figure 6.7 Elk Enjoying a Rancher's Green Grass in Paradise Valley, Montana. *Source*: Barbara K. Jones.

Deer on the roads and in yards generate a lot of unexpected costs to both drivers and homeowners, yet they aren't typically managed more effectively because deer, like elk in Montana, are tolerated in many places throughout New Jersey.

With reintroductions or rewildings, species recovery can add additional tension to the coexistence story. Although recovery indicates the rewilding or reintroduction effort has been a success, it also puts pressure on the community affected by the species' recovery. Even if we consider Auster's idea of renewed coexistence, people who have not lived with beavers, wolves, bears, or any other returning or recovering wild animal for generations might not be equipped to deal with the associated land-use restrictions. These restrictions could include beach, hiking trail, or ski slope closures, the return of wetlands to what was once a productive agricultural field, or the need to change their behaviors regarding wild animal attractants like bird feeders or household trash. Since these changes are imposed on people, they can make coexistence challenging. For ranchers and hunters who might never completely tolerate wolves, revisiting the human dimension of this reintroduction might have helped identify the incremental steps hunters and ranchers have made to coexist with a predator they have long hated. Identifying and highlighting these efforts is what a behavioral economist would call a nudge, where, in this case, the nudge rewards and incentivizes hoped-for behavioral changes that can lead to greater efforts at coexistence.

For beaver, as their numbers increase from natural recruitment, relocation, rewilding, or reintroduction efforts, education and increased awareness will help people shift their gaze away from beavers as a nuisance or a commodity to beavers as ecosystem engineers, as climate change mitigators, and as builders of natural resilience. That shift will change how we see public versus private property boundaries, how we manage species reintroductions and recoveries, and whether we can make strategies like relocations, beaver-dam analogs, painted trees, and beaver deceivers effective coexistence tools. Currently, our relationship with the beaver is on the ascent. In this time of climate change, the way beavers manage water through the processes of slowing, spreading, sinking, storing, and sharing makes it the animal to beat as we deal with drought, flooding, and the rebuilding of wetlands that are critical to both our well-being and nature's.

NOTES

1. Morgan 1868: 267.
2. Morgan 267.
3. Morgan viii.

4. Backhouse 54–55.
5. Goldfarb 12/14/2018
6. Krech III 194.
7. Castoreum is a yellowish product that comes from the beaver's castor sacs. It is used to mark territory.
8. HBC History Foundation, The Royal Charter.
9. Backhouse 123.
10. Fountain 242.
11. Shed hunting at the National Elk Refuge in Jackson, Wyoming is a major event each spring that occurs after the elk have dropped their antlers. Shed hunting is extremely popular, with massive numbers of vehicles lining up on opening day, but the local Boy Scouts have first access to the dropped sheds. Those they collect are then sold at the antler auction during the annual Elkfest Auction in Jackson, Wyoming. Twenty-five percent of the auction's proceeds help fund Scouting programs for boys and girls in Teton County and surrounding counties, while the remainder helps fund National Elk Refuge initiatives.
12. Morgan 246.
13. Backhouse 69.
14. Backhouse 69, 79.
15. HBC History Foundation. Maud Watt.
16. The Hoover and Shasta Dams would never have been built today. The cost of construction, as well as the loss of human life and species habitat, would have made projects like these impossible to consider as tools for flood control.
17. For the 2005 Millennium Ecosystem Assessment (MEA), the identified categories for ecosystem services include: provisioning services (goods produced by ecosystems—food, freshwater, fuel wood, and genetic resources), regulating services (benefits that come from regulating ecosystem processes—regulation of pollinators, erosion control, climate, diseases, nutrients, and extreme natural events), cultural services (the non-material benefits from ecosystems—spiritual, recreation, educational, knowledge, heritage), and supporting services (services necessary for the production of other ecosystem services—soil formation, flood control, nutrient cycling, carbon sequestration).
18. Methow Beaver Project.
19. Houck 7.
20. Jones. 2019. 18–19.
21. The Nature Conservancy.
22. Jones. 2019. 14–15, 17–18.
23. Greenaway 23.
24. Dickson 21.
25. Dickson 20.
26. Campbell et al. 24.
27. Grug.
28. Pinelands Adventures.
29. Hynes.
30. Haddock 10.

31. Goldfarb 2018: 79.

32. The Imperial Valley in Southern California is a blend of urban, suburban, and rural areas, with agriculture as the primary economic driver.

33. Fountain 245–250.

34. Fountain 251.

35. Hawkes, Alison June 19, 2014.

36. Lanman et al. 67; Lanman, et al. 195.

37. Lundquist and Dolman. 4. 2.

38. Haddock 13.

39. O'Neil 32.

40. Abrams 12.

41. Haddock 31.

42. Niemi and Fouty 4–5.

43. Sillars, Jordan September 24, 2021, THE BLUEPRINT FOR BANNING BEAVER TRAPPING.

MeatEater. The Blueprint for Banning Beaver Trapping | MeatEater Conservation (themeateater.com).

44. BARK.

45. Auster et al. 2021.

46. Auster email communication.

47. USDA APHIS.

48. van Zelst, unpublished, p. 1.

49. Perryman.

Chapter 7

The Question of Social Carrying Capacity

Unlike the biological carrying capacity that defines how many of a specific species a habitat can support, social carrying capacity is what our culture and perception tell us about the number of a species we can tolerate. This distinction is important when we consider species in recovery post-reintroductions or from the natural expansion of their range. The ESA, along with hunting rules and regulations, have allowed many species to recover, but that recovery can challenge our tolerance for coexisting with species that have long been absent or in such low numbers we have had little interaction with them. In some instances, the resulting recovery might be a biological success but not necessarily a social one. One-hundred beaver could be the biological carrying capacity of a particular wetland ecosystem, but if the community adjacent to this prime beaver habitat has a social carrying capacity of fifty, what happens to all those extra beavers? Lethal measures used to be the answer, but today we have many more options. We can relocate the surplus beavers, we can use pond levelers and beaver deceiver devices to minimize conflict, and we can work to increase the affected community's social tolerance. For beavers, these options are ways to minimize conflict and encourage greater coexistence.

By adding self-interested humans directly into an ecosystem, the ecosystem's resilience can weaken, threatening species' survivability within this altered habitat. To compensate for what can be our overbearing presence on the land, hunting regulations, species protection acts, habitat restorations, and reintroductions have become important tools to mitigate ecosystem damage or to enhance an ecosystem's natural resilience. Even with these tools, a challenge we face today is how to manage natural ecosystems when not only the biological and social needs of millions of different species across many different ecosystems must be met, but the social, economic, and ecological needs of the people who live within these same spaces must be satisfied as well. Some biologists and environmentalists have suggested a 30-by-30 approach, where 30 percent of both the Earth's land and water is protected, but to achieve that goal, cultures around the world must value and understand environmental protection in a somewhat similar way. The American "best idea" of national parks is another approach, but that again requires a shared understanding that national parks are spaces people visit but typically cannot occupy. Parks can also, by not including animal migration patterns, fail to effectively address the wildlife boundary problem. Marine management areas, no hunting or fishing zones, and closed beaches and wildlife areas are becoming more common as people encroach on bird and turtle nesting sites on beaches or alpine recreationists move into wild nature and disturb caribou, bighorn sheep, or mountain goats. This encroachment requires scientists and environmentalists to not only think about a species' ecosystem needs but also our social tolerance for recovering wildlife and the subsequent loss of access to popular landscapes.

CULTURAL HERITAGE AND INCONSISTENT MESSAGING

The failure to satisfy the overlapping and conflicting goals of species management is playing out across the United States, but no more acrimoniously than in the Intermountain West of the United States, where grizzly bears, elk, bison, and wolves are currently being managed through the lens of cultural heritage rather than science. A Montana rancher who lives in the Greater Yellowstone Ecosystem (GYE) north of YNP can be quick to tell you that he kills any wolf he thinks is "threatening" his cattle. What does threatening mean to him that justifies killing a wolf that might just be passing through his ranch? This rancher's low social tolerance for wolves is at odds with the Montana Department of Livestock's report that just seven head of cattle were killed by wolves between 2013 and 2021 in the Yellowstone region. Many more died from weather, disease, and feral dogs. Although agriculture is an important economic driver of Montana's economy, according to a study done by Wild Livelihoods, which is a coalition of over 150 local businesses in the GYE, wolf watching alone brings more than $82 million to the GYE communities each year.[1] For many ranchers and hunters, the economic force of wildlife viewing juxtaposed with the critical ecological function wolves provide has not significantly changed the potency of an anti-wolf cultural heritage as the way of seeing wolves on the landscape.

The loss of livestock from predation, whether it is cattle, sheep, or alpacas, can have a devastating emotional and ecological impact on the owner of any of these animals. Yet the reality is these ranchers and farmers operate in a landscape rich with wildlife, and to satisfy the environmental goals of a post-Rachel Carson era, a new relationship with the land and its native species needs to emerge. At a June 2024 meeting of the Idaho Wolf Depredation Control Board, a discussion on aerial shooting of wolves and concerns over Idaho's HB 612 that provides compensation for livestock losses without the presence of a carcass galvanized wolf supporters to speak. To one speaker, blaming wolves without a carcass furthers negative wolf stereotypes, while another pointed out how nonlethal wolf projects have been successful in other states but are rarely tried in Idaho. Instead, Idaho spends tax dollars to kill wolves, when the money could be used for nonlethal wolf-livestock coexistence strategies. Those comments provoked someone to ask how in other careers twenty-five years without changing skills is unheard of, yet somehow ranchers are exempt from adapting their skillsets "like everyone else." Just as a wolf cannot be allowed to consume with impunity a rancher's livestock, a rancher must be willing to adapt and learn to tolerate prairie dogs, predators, and bison on their land. By incorporating agrarian practices that more effectively accommodate wildlife's needs yet offer the rancher some agency with wildlife that competes with his labor should be a reasonable compromise in keeping with this post-Rachel Carson era.

Yet a low social tolerance for certain wildlife that is supported by varmint, pest, and vermin narratives continues to play out with species that are defined through a negative lens. Since coyotes have been called vermin, or in the 1920s the "Original Bolsheviks" or even fascists that threaten the American way of life,[2] they have been subjected to tremendous human pressure regardless of where they are or what they are doing. Recently, while driving up a public dirt road into a wildlife area near YNP, I saw a coyote walking across a large open field. While I stopped to watch the coyote, a white pickup truck raced past me and pulled off to the side of the road in front of my car. Within seconds, a young man was not only out of his vehicle with what apparently was a loaded rifle but was on his knee in position to shoot the coyote. The scene suggested not only a cultural heritage that sees driving with a loaded and ready rifle in your vehicle on a public road as acceptable, but a way of life that views coyotes with such disdain that a person would shoot the animal just because they could. It also presented a worldview that lacked any regard for someone who might assign more value to a live coyote than a dead one. Fortunately for the coyote, it disappeared into the sage before the man could take a shot. Dan Flores, in his book *Coyote America,* described the coyote as the most persecuted large mammal in American history. To him, how people feel about coyotes is such a powerful part of their identity that it is equivalent to religion, gender identity, and climate change as a hot-button cultural issue.

A similar situation to the coyote and the failed shooter occurred while visiting a prairie dog town at Greycliff Prairie Dog Town State Park in Montana. The prairie dogs in the state park are protected, but on the land immediately adjacent to the pocket-sized state park, they are not. As I walked back to my car at the park, I noticed several pickup trucks pull into a dirt lot next to the park boundary. From each of these vehicles, men with long guns piled out to set up for some target shooting. Their targets were the prairie dogs living outside the protection of the park. Like coyotes that, as vermin are disposable, prairie dogs as pests suffer a similar fate. Instances like these suggest that improving our social tolerance for these maligned species will require more than national parks and 30-by-30 land and water protection, but instead a reimagining of long-held "cultural truths" that have been told and maintained through storytelling and heritage-affirming language.

In a somewhat curious twist, hunters and anglers have created many "nuisance" problems with their own quest for game animals to hunt and fish to catch. In the 1960s and 1970s, to increase hunting opportunities, mountain goats were introduced to the Snake River Management Range near Grand Teton National Park (GTNP), while in Wyoming, non-native yellow perch were discovered in Saratoga Lake and walleye in Buffalo Bill Reservoir. The non-native goats migrated into GTNP and are now competing with native Rocky Mountain Bighorn, a species that is losing low-elevation habitat to development and is being outcompeted by mountain goats in higher elevations. In 2020, to rid the park of these goats, a cull was initiated, and by early 2022, one-hundred goats were removed by hunters with the ability and skill to hunt in very difficult mountain terrain. For Saratoga Lake, according to Bobby Compton, a Wyoming Game and Fish Regional fisheries supervisor, there is real concern that a fish illegally introduced by anglers looking for a more convenient spot to catch perch will threaten an important trout fishery because of the perch's prolific reproduction rates, while the walleye was introduced in Buffalo Bill Reservoir by anglers who wanted "the thrill of catching the feisty fish."[3] In each of these instances, an understanding of ecosystem health or what it really means to be a "nuisance" species did not dictate the hunter or angler's behavior; instead, a passion for fishing and hunting encouraged them to "manage" these ecosystems for their own purposes. By not understanding the important distinction between a non-native biological nuisance and a native social one, they failed to see any issue with their behavior.

Since hunting and fishing bring in a lot of money to states from licenses, taxes, and guided trips, state agencies can be complicit in creating new opportunities for hunters and anglers. Somewhat like the introduction of mountain goats in Wyoming, in Montana, a controversial "pheasant-planting" program sponsored by Montana FWP is being implemented on state lands to not only improve hunter recruitment and retention efforts by giving young hunters "more birds to shoot" but also to expand the wild non-native bird populations in the state. Since hunting licenses and taxes on firearms and ammunition provide a critical revenue stream for conservation, this effort is part of a strategy by state wildlife agencies around the country to improve not only hunter recruitment and retention but reactivation to hunting. With the plan to raise 50,000 birds, the "planting" is estimated to cost $1 million or $20 per bird. To some, this is not only a misuse of public funding but a risk of bringing disease into the wild population (which is not native but has learned to survive in Montana), and it does nothing for youth recruitment. According to Thomas Baumeister, Vice Chair of the Montana Chapter of Backcountry Hunters and Anglers and a past employee of FWP, providing young hunters with the opportunity to shoot farm-raised birds might give them a sense of success, but it won't provide them with a true hunting experience. In a letter to FWP regarding this effort, Baumeister wrote that "Presenting a fabricated put-and-take opportunity teaches to kill, rather than to hunt."[4] Of the 218 comments submitted to the FWP and published on their website, forty-four comments were in favor, with 174 in opposition. Those in opposition identified as their reasons Montana's stable "wild" population of pheasants, the risk from a highly pathogenic avian influenza (HPAI) that could contribute to future outbreaks in the wild, and the absence of any fair chase ethics, while those in favor were primarily dog training groups.[5] Marcus Strange, director of state policy and governmental relations for Montana Wildlife Federation, said, "The data and the anecdotal evidence suggest that this is not the most effective way to get kids into hunting . . . so why are we spending so much money to get birds into the field that don't add to genetic diversity of the population, that carry significant risk of disease transmission, and have really low survival rates?" Evidence supports Strange's point about low survival rates with a study from Idaho that demonstrated their wild female pheasants were seven times more likely to survive from spring to October than pen-reared females. In response to this evidence, Deb O'Neill, special programs director for FWP, explained that survivability of the pen-raised pheasants after release is not a high priority for FWP; instead, survivability is a bonus because "recruiting hunters, removing the barriers for young hunters so that they get interested and stay interested in hunting" is the actual point of the program.[6] An interesting mission for an agency in charge of managing Montana's wildlife and, like with beavers in Oregon, can lead to inconsistent messaging across agencies charged with wildlife management.

In each of these instances, a non-native species was initially not labeled as a pest, so no boundaries were imposed. The hunters got mountain goats or pheasants, while the anglers got perch and walleye. Yet today, where wildlife biologists in one part of a state agency are working hard to reintroduce native species and remove invasives, another part of the same agency is introducing non-native species as a pro-hunting marketing strategy. Since introductions of non-native species do little to expand our social tolerance for native species or increase our understanding of how ecosystems work, the need to supplement state fish and wildlife agencies with a new revenue stream becomes very apparent. Groups like WYldlife Tourism for Tomorrow recognize that and by creating a revenue stream to these agencies from wildlife tourism could shift how state wildlife agencies operate. Stocking non-native species on the landscape as the special programs of FWP's actual point could be replaced by better FWP messaging that includes improving habitat and, by educating all users about the value intact ecosystems provide us, increasing our social carrying capacity for native wildlife.

THE QUESTION OF SUCCESSFUL RECOVERIES

The recovery of sharks in Cape Cod and now in Long Island Sound is what many people would call good news in that their presence in the coastal waters of Cape Cod and New York are the result of positive human behavioral changes. Over time, we accepted the increased regulations on shark fishing, better management of their food sources from bunkers in the Long Island Sound to marine mammals (specifically seals) off the coast of Cape Cod, and we cleaned up our waters. These changes allowed species like sharks, bluefin tuna, whales, and dolphins to recover on their own. Regulations and better management are examples of nudges that incentivize people to change their behaviors. For generations, we lived in a world where seeing a shark in the wild or a seal on a public beach was unfamiliar and strange, but today the likelihood of seeing a wild animal in its natural setting is becoming more and more likely. Swimmers, kayakers, and boaters might see a shark or manatee in their shared space and need to modify their behavior to suit that encounter, while hikers and hunters might face a grizzly bear, so being prepared with bear spray can offer a better outcome for both. This is equally true for ranchers and farmers who now must contend with increasing numbers of wolves, grizzly bears, birds of prey, and other predators, as well as elk on their land. This increase in wildness requires a change in long-held behaviors better suited to a less domestic-dominated landscape. In each of these instances, accepting change is an ideal response to new conditions; however, the reality is most of us will operate within a spectrum of behavioral changes from those who refuse to change at all to those who welcome the changes with enthusiasm. The challenge is finding that sweet spot where most people can find a way to fit in with this new reality.

Although shark numbers are in decline globally, in New York Harbor, home to one of the busiest metropolitan areas in the world, shark numbers are on the upswing. That is environmentally a good thing, but in our imaginations, not so much. Since sharks range from the more common dogfish to the larger and more unusual great white, when people hear the word shark, they assume "man-eater" and panic. However, although sharks tend to be at the top of the marine food chain, attacks on humans are actually very rare. There is a one in 4.3 million chance of being killed by a shark.[7] Yet, while statistics tell us one thing, our brains and emotions tell us something else. In 2021, there were seventy-three unprovoked attacks from sharks globally, with the highest number of attacks occurring in the United States, specifically Florida. The rarity of shark attacks does little to mitigate our emotional response. We die from many other things a lot more often than a shark bite, yet for predators in recovery, messaging is critical because with more swimmers and sharks in the water, the chance of us and sharks engaging in behaviors that might lead to a poor human-shark encounter can only increase. Although, in percentage terms, according to Greg Metzger, chief field coordinator for the shark research and education program at the South Fork Natural History Museum in Bridgehampton, New York, "the chances of an 'encounter' are probably the same as they ever were: near zero."[8]

Typically, instead of educating us, shark messaging has tended to exacerbate our fears. The 1975 Steven Spielberg movie *Jaws* was premised on the idea that fear gets people into movie theaters and subsequently sells more advertisement time for programs like Discovery Channel's "Shark Week." What initially had been a Discovery Channel education program on shark biology and conservation morphed into a shark-fear show that led to sharks being portrayed as the bad guys in movies and even *National Geographic* programs relying on the theme "shark attack" as the viewer attractant. The Society of Environmental Journalists (SEJ) recognized that simply by watching television, people have been encouraged to think shark attacks are common, when according to The Diving Blog, you have a greater chance of winning an Academy Award in any given year than being bitten by a shark. In recognition of our risk to sharks rather than their risk to us, SEJ is encouraging journalists to rethink their storylines. Instead of the shark attack headline that is a fear tease, journalists should ask about ocean conservation and its relationship to shark conservation, what efforts are your local beaches making to allow people and sharks to safely coexist, or, to encourage consumer agency, what are the current laws in the United States regarding shark finning?[9] By relying on fear soundbites, "shark attacks" is the lens through which we relate to sharks, when in reality, we have been swimming with all sorts of sharks for years, and only because the emphasis has been on shark attacks rather than their resilience and ecological force as an apex predator have we been encouraged to fear them in such exaggerated ways. This fear has led in part to the killing globally of one-hundred million sharks per year for, in many instances, a bowl of shark fin soup.

Education can remind people that the ocean is the shark's home, not ours. Closing beaches, like lifeguards do after lightning strikes nearby, or flying drones over the water to spot sharks near bathers can ease people's fears, but better messaging that does not make the shark into a "man-eater" can be even more effective. People can learn to avoid swimming at dawn or dusk, in cloudy water, or near schools of fish. That kind of proactive messaging helps people move away from a "cull the sharks" to a "[their] house, [their] rules" perspective. That shift has allowed more New Yorkers at Robert Moses Beach on Fire Island in the summer of 2022 to adapt to the increased presence of sharks in their shared ocean.[10] In Seaside Park, New Jersey, after some teenagers kayaked one-hundred yards off the beach and caught a 200-pound sand shark, nj101.5 com (the online version of a popular New Jersey talk radio station) posted a picture of a great white shark with his toothy grin and the words, "Don't kid yourself . . . this is what you're swimming with at the NJ Shore." Their dramatic response led the article's readers to respond with comments that included "Sharks live in the ocean. Repeat after me, 'Sharks live in the ocean'" to the reality "watch for rip tides [because] many more people die of those than shark attacks at the Jersey Shore but that's not going to grab your advertising senses like a scary shark." This awareness suggests that coexistence is possible and is becoming more likely for species historically marginalized.

For gray wolves in the Intermountain West, their recovery has been amazing to behold. Yet it highlights, like with the shark, the role culture has played in forging human-wildlife relationships. The long-lasting enmity many hunters and ranchers have had for wolves has been simmering since the wolf's reintroduction almost thirty years ago. Hunters blame the wolves for what they claim is a decline in elk numbers, while ranchers blame the wolves for their livestock losses. This blame has allowed wolves to be managed by social acceptance rather than science, so instead of educating people to coexist with the animal, killing them is used as the perceived solution.

Although wolves have recovered successfully throughout the Intermountain West, neither side is satisfied. Wolf fans believe that an unreasonable number of wolves are getting killed in states that feel the animal is theirs to manage, while the state believes wolf numbers exceed what was mandated by reintroduction parameters. Accepting this reality, how do wildlife agencies shift how they manage these spaces so wolves that don't recognize national park land and state land as two distinct landscapes can coexist with wolf hunters and their *on-X* hunting apps who know the distinction exactly? The solution rests on the recognition that coexistence falls on a spectrum from some tolerance to full acceptance. Somewhere along that spectrum is a space that balances the wolf's role in maintaining both healthy ecosystems and positive human-wolf relationships with the negative messaging of an anti-wolf cultural heritage. Once that point is identified, both social acceptance of wolves and a tolerance for limited lethal management as part of a state's wolf management plan can become part of a New West reality.

SOCIAL AND ENVIRONMENTAL CONTRADICTIONS

Recently, I read in *Farm Progress*[11] that of the nearly three million cattle in Montana, approximately 240,000 of them are culled each year because they are infertile or past their reproductive years. These cattle are sent to market where they sell for negligible amounts because cattle that don't earn their keep are expensive to keep in the herd. By using the word culling instead of killing, the rancher's behavior is seen as a legitimate business decision somewhat removed from the culling process's actual endgame. There are two ways to understand the language of culling that can include the reduction in a wild animal population by selective slaughter or the sending of an inferior or surplus domestic animal to a slaughterhouse. Culling then is a process of animal removal that can be justified for economic, ecological, or social reasons. The cattle cull in Montana is obviously driven by economics, but how does culling cattle as a business decision reconcile with other cattle losses. Based on numbers from Montana's FWP, of those same approximately two to three million cattle, in 2022, severe winter weather killed approximately 27,000 while wolves, in 2023, were responsible for killing approximately twenty-three. If we think about the number of cattle culled because they were either inferior or surplus to be about eight percent of the total number of cattle in Montana (in 2022 there were 2,200, 000 head of cattle) each year and the number of cattle killed by wolves to be about 0.0010 percent of the total, it becomes apparent that cattle losses due to wolves are not the excessive economic burden described by ranchers. That leaves us with the social and emotional reasons to explain why wolves are seen as the enemy for many in the ranching community.

The cattle business is a challenging one, but living with nature in the West is part of its business model. Weather, disease, other predators, and toxic vegetation take out more cattle than wolves ever have. Yet the language of blood and guts across the landscape is the tried-and-true approach many ranchers use to maintain an open season on wolves. In a recent hearing about wolf-hunting quotas in Montana I attended, ranchers in favor of killing wolves talked about blood-soaked fields and the presence of scattered cattle bones, while those who saw wolves as an economic, social, and ecological necessity pushed back against the "sit, bait, and wait" philosophy that has defined the public's perception of Montana's new wolf-hunting laws. Instead, they argued that no other animal is treated like wolves, where decisions are made that "keep the [species'] numbers at its baseline." They reminded the FWP officials of the economics of wolf tourism, the negative side effects of wolf trapping to other species, including the protected grizzly bear, and they pointed out that since wildlife services kills any wolf that kills cattle why is an additional high quota hunting season even necessary. To them, doesn't Wildlife Services remove the "bad" wolves from the land? In response to the broad support wolves have across the country and Montana's political leaders feeling that wolves in Montana are the state's to manage, one woman said, "How would the American people feel if human effluent was pumped into Old Faithful [in YNP in Wyoming] or that paintballing was allowed at Roosevelt Arch [in YNP in Montana]? Those are world treasures like the wolves, and those actions would demonstrate the states were not acting as good stewards." Others focused on how wolves are nature's way of managing populations that includes controlling chronic wasting disease (CWD) in elk herds. Those supporting high wolf quotas or no quotas at all relied on Old West cultural heritage arguments tied to lifestyles that revolved around hunting, trapping, and ranching, while those opposed saw wolves through the lens of the New West, where wolves belong on the land and the services they provide to nature and our well-being have earned them the right to remain.

As vermin, though, the bar for wolves is very high. Their ecological and economic success as a recovering species is offset by a low social carrying capacity that is deeply rooted in history. The social intolerance for wildlife defined as pests or nuisances tempers the rules so that species like wolves, prairie dogs, beavers, and bison are treated differently than elk, deer, or moose. This difference in treatment allows hunters to operate outside fair chase and gives ranchers a free pass to overinflate a pest or vermin's impact on their economic bottom line. A low tolerance for wolves is a way of seeing the world that extends beyond wolves. It also includes an unwillingness to see other people's values or concerns through a more mutualistic lens or, as one speaker at the FWP Wolf hearing said, "my concerns are just as important as a trapper's." In today's American West, that attitude does not consistently exist. Despite relatively low numbers, some hunters, trappers, and ranchers have a tremendous influence over the stories that define species like wolves and grizzly bears. The stickiness of those stories emboldens the creation and passage of wildlife management policies that give preference to their worldview over any other. Efforts to shift the anti-wolf narrative will require these hunters to publicly acknowledge the value they place on the ethics of fair chase and legitimate use, and for ranchers to admit their livestock operations exist in a part of the country that is America's version of the Serengeti, where the survivability of wildlife and the success of long-distance migrations must be maintained at the scale of entire ecosystems.

Although as unpopular as wolves are culturally to some people, they are deadly to woodland caribou. The conflict between wolf recovery and the woodland caribou's decline, an animal that has disappeared from the lower forty-eight states and is seriously threatened in southern Canada, makes the wolf issue in the United States a challenging one to reconcile with caribou recovery. By living in our imaginations and not in our reality, the woodland caribou's presence on the land was overlooked, and so the effect of large numbers of predators on caribou survival was not foreseen by scientists. Just as the fungibility of the animal in our imagination allowed moose, elk, and white-tailed deer to substitute for caribou, their status as ghosts has made ever seeing a caribou unlikely, so when they blinked out, people didn't even notice. The way they move up and down the landscape is critical to their survival, and by maintaining old-growth boreal forests that provide little food for predators or other browsers let woodland caribou evolve in an ecosystem with little dietary competition and a limited number of predators. They could travel in small herds, live off lichens, and calve where deep snow acted as a predator barrier. These behaviors worked well in a world where human actions hadn't altered their habitat and access to food sources or led to increased competition from other browsers that, through a preference for white-tailed deer, led to the unintended consequences of increased predation on caribou.

Wolf reintroduction in YNP and their subsequent recovery throughout the Intermountain West and southern Canada and a changing public perception about wolves has made caribou recovery extremely challenging. It also has forced wildlife managers and caribou conservation groups to make decisions about protecting one animal at the expense of another. For caribou in Canada, the feeling among those committed to caribou recovery is wolves will recover in Canada even with hunting pressure, while caribou will not survive without reducing wolf numbers. By overlooking woodland caribou when we had a chance to save them, recovery today faces significant odds. Odds that require those vested in caribou recovery to rely on maternity pens, conservation breeding, and translocations, as well as the need to prioritize caribou over wolves in caribou habitat. To a Kalispel Tribe wildlife biologist, the money it would have taken to keep caribou from disappearing was "budget dust," and yet for those living in caribou habitat was impossible to find. Caribou survived for thousands of years coexisting with predators, but over the past eighty years, human actions have put unnaturally high predator numbers, as well as loggers and alpine recreationists in caribou habitat pushing the animal to its tipping point. To make this recovery real will not only require improved habitat but also a willingness on our part to maintain boreal forests that provide a nutrient-poor barrier to wolf expansion into prime caribou habitat.

CHANGING STORIES TO CREATE PRAGMATIC SOLUTIONS

Shifting wildlife stories is challenging because they are rooted in worldviews acquired through lived experience or are socially reproduced through generations. Since time and a shared understanding of the world make cultural heritage arguments a powerful force in keeping these stories alive and meaningful, living as an outlier within these communities can be very difficult. Yet for those who do, there can be surprising results. One rancher in Montana set in motion a process of turning culled cattle into food for Montanans in need. Many of the same ranchers who cull their nonproductive cattle for business reasons are now willingly sending some of these animals to a nonprofit called Producer Partnership. That nonprofit provides the Montana Food Bank Network with free hamburgers, which is then distributed to food banks around the state. This is an example of ranchers serving as problem solvers by blending the need to cull cattle that have little market value but have food value with the need to provide hungry people in their state with food. From this nonprofit, Montanans in need get high-quality, low-fat hamburger, the rancher gets a tax deduction for a charitable donation for the 350 pounds of meat each animal provides, and community members can receive job training in meat processing and butchering through internships at the butchering plant or through a meat processing program at Miles Community College in Miles City, Montana. So instead of a zero-sum game, this solution to a problem generated a tripartite of winners.

The ability to find solutions to problems requires identifying the problem and then finding solutions that offer common ground to competing stakeholders. For predators, the problem is coexistence, and the solution will require answers that allow the largest number of individuals some satisfaction with the outcome. Killing wolves just because a person can might appease a few hunters and ranchers in the short-term, but it alienates a large constituency of people who see value in predators, like wolves, and want them around. The optics of high or no wolf-hunting quotas, political arguments that "the people of Montana don't want wolves," or bumper stickers that read "smoke three packs a day" appeal to one small subset and are not seen as a solution for anyone but a few powerful ranchers, hunters, or trappers. Shifting how they see wolves is one critical step to increasing social tolerance for them as a species. To Chris Servheen, position discussions are not helpful in finding common ground, but interest discussions are. A rancher who doesn't want to lose livestock to grizzlies has a problem that can be addressed with conflict management practices, but comments like "I hate bears" or "We don't want the federal government telling us what to do" are positions that "are worthless because," to Servheen, "you end up hitting a wall."[12] Cultural positions cannot only be deeply entrenched but can also trip up efforts at pragmatic solutions. A better approach might be to encourage these hunters, trappers, and ranchers to see that many people living within their own communities want wolves around. Accepting that other points of view have value is what community citizenship is all about (see figure 7.1).

Although wolves have had a long history as vermin and their absence gave many a sense of what life could be without them, their recovery along with that of bison, grizzly bears, bald eagles, and beavers have forced us to ask how to increase the social carrying capacity for species long gone that are now returning in significant numbers. Rebranding is a critical tool because it allows us to rethink our labels, stories, and descriptions to ones that see wildlife as our neighbors on a shared landscape. Rather than be shocked that a black bear is in your neighborhood, or a rare Florida panther just ran in front of your car, by normalizing them as members of our communities we might be better equipped to live with them in harmony. White-tailed deer in New Jersey challenge the homeowner and gardener with their love of lawns, plants, and food crops but any effort to reduce deer numbers through sharpshooters or birth control is not typically well received. Even though thousands of deer are killed on the state's highways each year, they, through a history of good branding, are extremely popular. Like in the Intermountain West, so deer and other New Jersey wildlife aren't destined to become roadkill, wildlife crossings have become part of the discussion on managing deer and other species in this highly urban state. Connecting Habitat Across New Jersey (CHANJ) is a tool the New Jersey Department of Environmental Protection is using to identify corridors and to find partners to achieve wildlife protection goals.

Figure 7.1 Cattle Drive in Paradise Valley, Montana. May 2022. *Source*: Barbara K. Jones.

Wildlife corridors and crossings are tools that have popular support across the country and remind us of the social intolerance-boundary issue. By having wildlife travel under or over roads and through tunnels, it becomes possible for them to move through our shared spaces with minimal negative impact.

Across the country, wildlife crossings tied to wildlife corridors are having a significant effect on many species' survival. In the American West, private landowners are helping wildlife move freely along traditional migration pathways, while in Florida, wildlife corridors are creating greater connectivity across an increasingly developed landscape. These corridors help panthers expand their range, bears move safely through human-managed landscapes, and provide critical habitat for whooping cranes, wood storks, West Indian manatees, and snail kites. The Florida Wildlife Corridor encompasses 17.7 million acres that include 9.6 million acres of protected land and an additional 8.1 million acres of opportunity land that is currently not protected but has the potential to be. Of particular interest is that seven million acres of the 17.1 million that make up Florida's Wildlife Corridor is working timber and ranchland, while of the 8.1 million acres of opportunity land, 33 percent is ranchland and 43 percent timberland. The ownership breakdown is even more interesting in that 4.2 million acres are federal land, 5.7 million acres are state land, 518 thousand acres are county and city land, and 311 thousand acres are private land with permanent conservation status.[13] The success of the corridor is based on the incorporation of working lands into the conservation plan. A similar model of conservation works in the Galapagos Islands, where the famous Galapagos giant tortoises can live on cattle ranches to the benefit of both the tortoise and the rancher. The rancher can not only supplement his income by allowing tortoise viewers to wander his ranch but can also benefit from the tortoises' role as ecosystem architects, while the tortoises benefit from the habitat provided. Coexistence comes in all forms, but because politicians and wildlife agencies reflecting the will of the people see value in Florida's wildlife, a greater effort is being made to encourage an increased social tolerance for Florida's native species and landscapes.

This ability to incorporate working lands into a conservation model is critical yet can be unpopular in many parts of the country. The nonprofit American Prairie (APR) applied in 2018 to obtain grazing leases from the Bureau of Land Management (BLM) for the purpose of grazing bison in Phillips County, Montana. APR's goal to conserve prairie species by cobbling together large public parcels and connecting them with private land is to create the largest wildlife reserve of its kind in the lower forty-eight states,[14] and by relying on mostly private money, it has been successful in buying grazing leases and available ranchland. During the public comment period regarding these leases, the BLM received over 2700 comments about this highly controversial plan of returning bison to land in Montana. Those opposed tended to be limited to Montana's Republican governor and legislature and some industry groups, with Montana Attorney General Austin Knudsen claiming, "it's not a surprise that Biden's Bureau of Land Management would rubber stamp this radical proposal that is another step toward displacing northeast Montana's livestock industry and replacing it with a large zoo." Governor Gianforte questioned whether the BLM had "statutory authority," and Senator Steve Daines called the approval "completely unreasonable" and an example of how the Biden administration "ignore[ed] input from [Montana] ranchers.[15] Curiously, a conservative market-based environmental group called Property and Environmental Research Center (PERC) had a totally different take on this effort. Hannah Downey, PERC policy director, presented a public comment to the BLM in support of APR's request to graze bison instead of cattle on a federal grazing allotment. She said:

> Contrary to the lobbying and litigation model that can make conservation a source of conflict, APR seeks to establish a large nature reserve by purchasing private lands from willing sellers, operating grazing leases on neighboring public lands, and incentivizing neighbors to adopt wildlife- friendly practices. As PERC's research has shown, federal grazing leases too often exacerbate conflict and APR's alternative market- and property-based approach can promote better conservation outcomes. For these reasons, we support Alternative B[16] to allow APR to convert the public grazing permits from cattle to bison.[17]

In this summary, Downey highlights how working with the community by buying land from willing sellers, by using neighboring public lands to graze bison, and by incentivizing neighbors to be more wildlife-friendly can lead to greater success with conservation initiatives. Incentivizing new relationships with wildlife is an extremely important piece of this because it reminds us of how critical it is to change deeply entrenched stories and worldviews. By relying on willing sellers and incentives that reward wildlife-friendly ranching practices, shifting the Old West narrative based on domination over nature to the New West narrative of conservation of wildlife makes more sense than the optics of kill what can be killed, and the economics of both ranching cattle in an era of climate change and a decline in consumer demand for beef. For ranchers in Montana's northeast, the market does not explain why they stay in ranching, but it does explain why there are so many willing sellers. Combining public and private land into a protected but working landscape offers a path forward for our diminishing prairie ecosystems. It also provides new opportunities for ranchers in Montana's northeast and, by responding to changes in consumer demands regarding beef and nature-based activities, a way to revitalize the social, ecological, and economic vitality of communities in the region.

When Alison Fox, CEO of American Prairie (APR), asked Kris Tompkins, president and co-founder of Tompkins Conservation, in an interview for American Prairie's *Sentinel* publication, "What are some of the universal truths you've seen in these large-scale conservation efforts?," Tompkins responded by saying,

> First and foremost, wherever it is in the world, there is a natural conflict between conservation and production . . . [to be successful with conservation efforts] takes put[ting] in the years and be[ing] a good neighbor and finding a way that all of this works and is worthy of local and regional communities [because] they will be the ones who protect the place for the next 100, 150, 200 years.[18]

This interview further highlights the value of building relationships that encourage coexistence through compromise and goodwill.

The state government's concern over American Prairie's goal of shifting land out of agricultural production to conservation or recreation use also fits squarely with an Old West worldview. Not only is that view tied to domination over nature and old notions of masculinity, but to productive land use practices as well. As an alternative, working lands offer an outlet for production through the lens of coexistence, rather than through domination over nature. For many of today's nature advocates, the tenets of coexistence more closely align with new economic and ecological realities, as well as ongoing changes in social values.

For the BLM, the accusation by the Montana Public Lands Council, an agriculture industry group, that the BLM gave "preferential treatment to this specific permittee" fell flat because the BLM has permitted bison grazing on other BLM allotments both in- and out-of-state. The record of decision from the BLM stated:

> Though the proposal to allow domestic indigenous livestock grazing conflicts with views and opinions expressed among some users of public lands, such unfavorable views of the proposal *do not constitute a scientific controversy*, disagreement about the nature of effects, or provide evidence that the project is not in conformance to BLM's statutory and regulatory requirements.[19] (Italics my own)

By tying their response to "domestic indigenous livestock," where APR will "own" the bison, the BLM could use the bison-as-livestock metaphor to make their case. The inability to see bison as a wild animal in Montana has made this decision by the BLM much easier. Since bison have been managed and treated as livestock across the state, leaving no truly wild herds outside of YNP, the BLM had no obligation to treat bison any differently than they would cattle on a grazing permit. Treating wolves as vermin and bison as livestock have put the state of Montana in the awkward position of explaining how vermin can generate millions of dollars in tourist revenue, manage ungulate populations, and restore ecosystems, and how livestock is only livestock when it doesn't interfere with rancher access to grazing leases. What then are bison if they aren't wild, yet aren't livestock either? Poorly defined labels enable inconsistent yet rigid and impractical boundaries to not only persist, but to maintain a separation between human and wildlife communities that Aldo Leopold considered to be the challenge of our time to correct.

The idea that wildlife stories, metaphors, and descriptions built by different settler streams create boundaries for wildlife plays out around the country, but probably nowhere as rigidly as where the Far West's relationship with nature still defines the landscape. In many other parts of the United States that include what Colin Woodard described as settler streams of Yankeedom, New Netherlands, the Midlands, Greater Appalachia, and the Left Coast, the transition to a more mutualistic relationship with wildlife is underway. The pleasure of seeing peregrine falcons, whales, or snowy owls in and around New York City or black bears wandering downtown Asheville, North Carolina, or Gatlinburg, Tennessee reflects a growing tolerance for wildlife in our human spaces. Manatees in Florida are also evidence of a transition to mutualism. Since they live in the coastal zones and inland waterways, manatees rely on the willingness of many boaters and swimmers to share bodies of water with an animal that until the 1970s was practically unknown. In areas where hunting, trapping, fishing, unrestricted recreation, and farming are highly valued cultural activities, the tolerance for wildlife recovery can meet with more resistance. Even the manatees in Florida need to be managed as well as protected by friends of manatees called manatee heroes. Some boaters still see manatees as an intrusive pest, so slowing down in a no-wake manatee zone is something they choose not to do, while pesticide and fertilizer run-off contribute to poor water quality, making it difficult for manatees to find food. During the winter of 2021–2022, over one thousand manatees, or approximately ten percent of the entire population of Florida manatees, died on the state's east coast from starvation and cold. After such a significant mortality event, legal pressure and a successful lawsuit from concerned environmental groups forced the USFWS to commit to publish proposed revisions of critical habitat protections by September 2024. For those involved in the lawsuit against the USFWS, they saw the manatees in Florida as the "canary in the coalmine when everyone's [from sea turtles getting tumors to fish losing their mangrove habitat to a decline in commercial and recreational fisheries] house is on fire."[20] The conflict between development and protection over land and water access in Florida has allowed the USFWS and the EPA to let manatee protections from 1976 remain unchanged. Changes in how society deals with farms and residential run-off, and the habits of some recreational boaters and swimmers will only occur if manatees are recognized by most stakeholders as a valued part of Florida's landscape. After the recent significant manatee mortality events, the stringent protections manatees had while listed as an endangered species might return as the USFWS reevaluates their status as threatened rather than endangered.

For those species living within a prairie habitat, the cultural heritage of using prairie dogs as target practice was and continues to be justified because prairie dogs are varmints or nuisances that need to be eradicated by either a gun or poison. As varmints, they are not often welcome on land where cattle graze, so they are perceived as disposable. When asked why anyone would shoot prairie dogs, one hunter said it was a "cultural thing" that helps ranchers and farmers who have problems with varmints like prairie dogs, woodchucks, or coyotes. This hunter also believed shooting small live targets makes for a better hunter with a more proficient shot. In response to a YouTube video called "Prairie Dog Armageddon 4: Why Shoot Prairie Dogs?," viewer comments ranged from "absolute fun in a bloody guts way" to "every day during spring and summer as a kid sighting my rifle on those bastards preparing for deer/elk season," to "I just got home from doing this in Montana for 2 weeks. It was so fun. The ranchers were so glad we were there." Another prairie dog hunter wrote in a blog post called "Prairie Dog Plinking—Great Summertime Rifle Practice" that as a "summer activity . . . [shooting prairie dogs is] not only fun but help[s] others" by reducing the spread of bubonic plague and the number of prairie dog holes in a cattle pasture.[21] To clarify how the killing of prairie dogs is a summertime fun theme, plinking means informal target shooting done for pleasure. The more typical "victim" of plinking, though, is a roadside sign or a soda can, which to this blogger is apparently the moral equivalent to shooting a prairie dog. On Dan Cattaneo's Predator Control Varmint Hunt webpage, he suggests bringing a second gun when going on a prairie dog "hunt" because there will occasionally be so much shooting that your first gun gets too hot. Attitudes like these require an answer to how "blowing up" prairie dogs can honor the ethic of fair chase or legitimate use? It also begs the question whether those engaged in this cultural practice ever consider changes in social mores as a legitimate reason to create a substitute for this traditional summer activity.

To avoid the ethical dilemma associated with killing a prairie dog just because it is possible to do is to rationalize that a dead varmint has more value than a live one ever could. By shooting the animal so that it explodes in a blur of blood and fur, its death offers the target shooter something of value and that is the maintenance of a way of life. Like the bison and brucellosis argument, prairie dog killing is reliant in part on misinformation that the prairie dog is a carrier of the bubonic plague, when it is not a reservoir for the disease but a victim of it. When a prairie dog town gets the plague, up to 90 percent of its members typically die, so, through a quick death, they are less likely to contribute to the plague's spread. This adversarial relationship with prairie dogs is also encouraged by laws that allow the species to be killed at any time without a hunting license. In Wyoming, there are few restrictions on killing prairie dogs because they are considered a pest species. Yet today, both scientific evidence and anecdotal observations from the actions of bison show that prairie dogs provide ecosystem services that support hundreds of other species, as well as improve grass quality, and through their gardening habits recycle nutrients and nourish other plants. Since bison have coevolved with prairie dogs, they seek out the grass near prairie dog burrows because it is tastier and more nutritious. This growing awareness about the role prairie dogs play in maintaining grassland ecosystems can make the animal more welcome on some ranchland, but it has done little to stave off the tradition of live target shooting. For prairie dogs, the language of varmint and nuisance has maintained a low social tolerance for their presence on the land.

CAN COEXISTENCE BE THE SOLUTION?

Accepting that language reaffirms a specific way of seeing the world, it becomes possible to use language to reimagine new ways of seeing and interacting with the natural world. The soundbites that created the beaver-as-a-pest metaphor are being flipped to beaver as ecosystem engineers. Promoters of beaver have discovered that it is possible to rebrand them in ways that encourage coexistence. The Smokey the Beaver mascot is an example of imagery that is readily understood and, through familiarity, is popular with people. With that approach, the Beaver Institute promotes beaver coexistence as "a good idea" that, through good marketing, increases social tolerance for beavers. A recent public awareness campaign relied on the idea of "Beaver Biobanks" as a catchy and easily understood way to generate media interest, encourage public support, motivate funders, and inform policymakers.[22] The idea is pithy and, by relying on the double meaning of "banks," the language of "Beaver Biobanks" can suggest to people that beavers create landscapes as well as banks of natural capital. The campaign further relies on the banking metaphor to encourage building partnerships and "branch offices" to promote nonlethal beaver coexistence, establish new beaver wetlands, train BeaverCorps wetland professionals, and build climate resilience. By creating a National Beaver BioBank Program to spread the beaver message, the role of beavers in building climate resilience and improved economic opportunity can be more readily understood in communities throughout the United States.[23] Such positive messaging makes human-beaver coexistence easier to achieve.

Flipping the message has worked with other species as well. In the Lowcountry of South Carolina in Beaufort County, alligator numbers have grown tremendously since the 1960s when a poorly regulated "harvest" resulted in a severe decline in alligator numbers. Today the population has stabilized, but after a few members of a Sea Pine Island community in Beaufort County died from alligator attacks, there have been conversations about how to safely coexist with this apex predator. Fortunately for the alligators, the residents of this community know they chose to live where alligators, not humans, are at the top of the food chain. That reality is framed in the notion that the land they have turned into a gated community for retirees used to be a swamp full of alligators. Because of that, people seem more willing to learn to live with them and to educate themselves about alligator awareness and safety. Like with wildlife anywhere, an important message they share is "a fed gator is a dead gator," making the availability of anthropogenic food to alligators something no one should do. Their coexistence efforts also include keeping four car lengths from an alligator when walking, keeping pets and children away from the water's edge, not feeding or harassing alligators, and, for anglers, disposing of their used bait away from the water. The feeling by many in this community is that education is better than euthanizing large numbers of alligators.[24] Even with two deaths from alligator attacks in recent years, the community remains committed to maintaining a high social tolerance for alligators, which has resulted in a willingness to coexist with them.

Today, coexistence comes in many forms. A landowner who modifies his fences for pronghorns or a rancher who welcomes the return of beavers because beaver pools provide his cattle with water are both examples of coexistence. Even fruit gleaning is a public effort at sharing space with bears. A Montana group on Facebook called Flathead Fruit Gleaning allows people with fruit trees to post fruit availability, and by inviting people to pick their fruit trees in the early fall before the bears arrive during hyperphagia, the bears become less reliant on fruit in yards or orchards. This more accommodating approach to wildlife as diverse as beavers, pronghorns, and bears has taken time and a significant amount of effort. Yet, according to a Nevada rancher interviewed for a September 6, 2022, *New York Times* front-page article about beavers, "[Beavers] are very controversial still. . . . But it is getting better. People are starting to wake up."[25]

For elk, social tolerance is typically so high that no one even uses the word coexistence to describe the relationship between landowners and elk. Even though elk have the greatest dietary overlap with cattle and are more likely to transmit brucellosis to cattle than bison, ranchers don't require an elk image makeover or the potential for cash payments to see elk as part of the landscape. Yet by using the word coexistence for those species that challenge many people's tolerance, the word becomes conflated with the species and can trigger a negative response. Malou Anderson-Ramirez avoids using coexistence when talking to ranchers because she believes the language of coexistence is linear when the behaviors coexistence is trying to encourage are circular. To her, coexistence can't be either/or; instead, it needs to capture the gray space where people are. That means when you are trying to learn about a rancher's approach to cattle management, it is better to ask, "How do you keep your cattle alive?" rather than "How do you keep wolves out?" To promote the ideas of coexistence, Anderson-Ramirez prefers to meet people in that more nuanced gray space of ranch management rather than the hotly contested space of wolf control.

Whether the Beaver Biobank message turns you into a beaver believer or you begin to welcome predators on your land because predator-friendly branding earns your beef premium pricing, you are engaging in coexistence. Today's regenerative agriculture is circular, just as rewilding is. So when people add beavers, share space with woodland caribou, remove fences, or accept small economic and emotional losses to maintain healthy predator populations, they are circling in and out of those gray spaces. Through that process, they can begin to see that othering wildlife with labels like vermin and varmint becomes illogical when the varmint improves grass quality for their cattle and the vermin brings in millions of tourist dollars that provide local job opportunities for their non-ranching children. Incremental change that brings the entire planet forward might seem painfully slow, but as we reimagine wildlife through a new lens, we can begin to remove those imaginary boundaries that negative wildlife metaphors and stories have created and maintained. Porous boundaries along with better stories can only help increase our social tolerance for species typically marginalized so that coexistence rather than human-wildlife separation is something more of us can support.

NOTES

1. Wild Livelihoods.
2. Flores 19.
3. England.
4. Collins, July 23, 2022.
5. Wyoming's pheasant-stocking program had to euthanize 1,200 breeding-age pheasants after they were exposed to HPAI by wild turkeys that died from the disease near one of the breeding facilities. Hall, 6/16/22.
6. Ibid.
7. Wildlife Museum.
8. Helmore July 26, 2022.
9. Davis. July 27, 2022.

10. Ibid.
11. Thomas. February 3, 2022.
12. Deveraux January 29, 2023.
13. Florida Wildlife Corridor Foundation.
14. Eggert. July 29, 2022.
15. Ibid.
16. Alternative B is APR's revised proposal in response to public concerns related to bison year-long continuous grazing. It identifies APR's expected bison stocking plans for the next ten years or more. It is based on APR's intent to keep bison management operations centralized and to have herds of at least four-hundred animals to maintain genetic diversity. Environmental Assessment. American Prairie Reserve Change of Use DOI-BLM-L010-2018-007-EA. June 2021. Section 2.6.
17. Downey, July 22, 2021.
18. Sentinel. 4.
19. Eggert. Italics my own.
20. Gomez-Upequi. June 10, 2022.
21. Bachman.
22. Beaver Institute.
23. Beaver Institute.
24. Haselhorst and McDonald, August 30, 2022.
25. Einhorn A1, A15.

Conclusion

As this book comes to an end, the hope of Aldo Leopold that humans, animals, and plants would be understood as one mutualistic community going about the business of living remains unfulfilled. In February 2024, a young wolf that was run down by a snowmobiler, transported, and then killed in Wyoming became international news in part because the alleged perpetrator's punishment was only $250. In a state where torturing a dog or goat can result in up to two years in prison and a $5,000 fine, a minuscule fine and no prison time for engaging in the same actions on a wild wolf is indicative of how public officials see this animal. In an area southeast of Wyoming's Grand Teton National Park, running down wolves with snowmobiles or shooting them from airplanes has long been accepted behavior, so the horror and shock many Wyomingites have expressed over this incident rings a bit hollow. Voters have continued to elect officials who encourage policies that allow for a set of behaviors most people would see as both cruel and evil. Such behaviors do not occur in a vacuum. Instead, they come from generations of wolf intolerance and a society that treats wolves as vermin across most of the state.

In another recent incident, after leaving its winter den in the spring of 2024, a young grizzly bear made its way into the Bighorn Mountains of Wyoming. In a state where officials are eager to get the grizzly bear delisted, a subadult grizzly traveled eighty miles east of Wyoming's Demographic Monitoring Area (DMA), an area considered to be biologically and *socially* suitable for grizzly bears, for the first time in nearly one-hundred years. After wounding a cow that was later euthanized due to its injuries, in conversation with the USFWS, the Wyoming Game and Fish Department then killed the young bear. According to Game and Fish Director Brian Nesvik, "The Bighorn Mountain Range is not suitable habitat, and the department is *not* interested in allowing grizzly bears to occupy this area." Due to potential human-grizzly conflict, the department believes the bears' expansion into unsuitable habitat is an impediment to the success of grizzly bear conservation.[1] Yet if the goal in the state is to improve grizzly bear connectivity for eventual delisting, killing a young bear in what Wyoming Game and Fish believes to be unsuitable habitat only makes it more difficult for the species to expand its range. Its death also does nothing to increase Wyoming's social carrying capacity for grizzly bears.

This low social tolerance for grizzlies fits in with Chris Servheen's comments about Montana's Fish, Wildlife, and Parks' draft grizzly management plan post-delisting. To him, this attitude feeds into existing "anti-predator paranoia" and is written with a "tone of intolerance."[2] For an animal that at one time ranged in numbers that exceeded 50,000 from the Pacific Coast to the Mississippi River to Central Mexico, the draft plan with its limits on bear movement and an endgame of trophy hunting does nothing to increase social tolerance or make predator boundaries more porous. Instead, it offers another example of how wildlife management solutions for those species that challenge us and forget their "place" tend to be political rather than scientific. When a beaver reintroduction participant in the United Kingdom said, "because of the way beavers use the landscape [. . .] it should be a catchment scale [. . .] It should be based on geographical boundaries, rather than political boundaries,"[3] the participant could have replaced beavers with grizzly bears, wolves, caribou, or bison and the argument about management scale would have remained the same. In 2017, the presence of an estimated 700 bears and a doubling of their 1970s range in the GYE led then-Interior Secretary Ryan Zinke to announce the delisting of Yellowstone grizzlies. To him, "[t]he grizzly bear recovery] stands as one of America's great conservation successes." The argument for delisting was and continues to be based on the idea that at approximately 1,100 bears occupying four percent of their original range is "recovered."[4]

To Mark Heinz, the reporter for Wyoming's *Cowboy State Daily* who covered the bulk of the Wyoming 2024 wolf torture story, the grizzly bear used to be the unpopular kid on the block, but once the wolf was returned, an animal far less liked, the grizzly bear didn't look so bad. The social tolerance in Wyoming for grizzly bears is higher than that for wolves because the feeling is grizzly bears are not going to "decimate a herd," so they are not seen as an "existential threat like the wolf." Heinz believes the social tolerance for both grizzlies and wolves, with the tolerance for wolves being significantly less than it is for bears, is weakened by the perception that the federal government "forced" these animals upon the local population. The wolves get the brunt of their anger because, while the grizzlies were placed under federal protection and allowed to recover on their own, the wolves were deliberately brought in and reintroduced in 1995. When Dr. Roger Auster commented on how there exists a "link" between the reintroduced beavers and "the people [who] put them there," he could just as easily have been talking about the gray wolf.

Although the social tolerance for grizzly bears might be greater than that of wolves, for both species there is an expectation that they will be intensely managed. The nature of this management has allowed endangered grizzly bears to be killed as a first resort (instead of using bear spray, making noise, traveling in groups, or avoiding carcasses) rather than as a last resort with no other option. Wolves, in turn, can be killed at almost any time or for any reason as predators in 85 percent of the state. In Wyoming, since predator species also include porcupines and jackrabbits, predator is just another word for vermin or pest.

For wolves, negative attitudes are deeply rooted in the past for many Wyomingites, with the outrage around them simmering until the wolf reintroduction in 1995 to Yellowstone National Park (YNP). In those small communities around the state that never really lost their anti-wolf mentality, the reintroduced wolf became the face of what Heinz called "big-city folk and federal government encroachment." Moreover, labeling these wolves "supersized invasives" helped intensify the feeling that groups living outside of Wyoming were making decisions for them. From this, the Wyoming Game and Fish established legal predator (vermin) zones for killing species labeled as predators. This atmosphere of managing wolves as predators or vermin in Wyoming was an accepted and established approach until that moment in February 2024, when a "hunter" ran down a juvenile wolf with his snowmobile, disabling the animal. He then brought the wounded animal home, wrapped its muzzle with red tape, and transported it to a local bar for show and tell. After leaving the wolf suffering on the bar floor for a time, the man took the wolf outside and killed it. The public knew nothing about the event until a month later when pictures were leaked to the press. Since then, the outcry regarding this hunter's behavior and the token punishment he received, a small fine for possessing and transporting a live wild animal, has been explosive. To Heinz, in all his years as a reporter, this story has taken off and spread globally in a way he has never seen before. As a lifelong resident of the Rocky Mountain West who has lived in Wyoming for twenty years, the killing of wolves is something he recognizes rural Westerners do, but taping up a wolf's mouth and displaying it publicly does not match what even the most dedicated predator hunters would consider ethical behavior. Most hunters agree that an animal should be killed quickly, not deliberately made to suffer for a prolonged period. This leads him to wonder if the intense public response will "homogenize [Wyomingites] as 'knuckle-draggers' who torture wild animals for fun." Already he has been told people are canceling vacations or their plans to move to Wyoming, suggesting this incident might be "the zeitgeist moment" for wolves. This could have been the event that not only increased Wyoming's social tolerance for wolves, but also changed how they were managed. Yet, in July 2024, after months of discussion, some members of Wyoming's legislature were able to rationalize that killing wolves by whacking them with snowmobiles was more humane than other lethal methods. Such a rationale indicates this will not be the moment of change for Wyoming's gray wolves.

WILDLIFE DISCOURSE

The language of wildlife discourse is fraught with human subjectivity and its ties to cultural tradition, experience, and individual worldviews. For science to function, we can't as Dr. Carlo Rovelli, a theoretical physicist and author of *Anaximander and the Birth of Science*, wrote "zero the tangle of our assumptions" that comes with our very humanness, but we can free knowledge through reliance on scientific reasoning from what he calls "the prison of tradition."[5] In the case of grizzly bears and wolves, the respective management plans incorporate a tradition of intolerance for predators as a preconception into how these plans are designed. For grizzlies in Montana, those that travel east of the Northern Continental Divide Ecosystem (NCDE) and GYE will be "tolerated only insofar as they remain conflict free" and that their "presence would not be an objective in areas far from their largely mountain habitats and prairie habitats where agricultural development predominate."[6] A reliance on human-defined "recovery zones," which David Mattson of *Grizzly Times* at a talk sponsored by Wildlife Guardians titled "Grizzly Bears in the Northern Rockies," called "artifacts of human ideas," makes the natural expansion of grizzly bears less likely because the people living outside of those zones are not encouraged or expected to accommodate Montana's official state animal on their landscapes. In most instances, the unexpected nature of a grizzly bear in the northeast or central parts of Montana, or even between recovery zones, can result in the bear's death. According to Kristin Combs of Wyoming Wildlife Advocates, at the same talk as Mattson, people are unprepared for the arrival of grizzly bears in part because they believe grizzly bears are supposed to live in high mountain forests when they evolved on the plains. Yet for all that, in 1990, the recovery zones grizzly bears could occupy consisted of approximately 9,200 acres around Yellowstone and Grand Teton National Parks, while today grizzly bears can potentially occupy 42,000 square miles of land in Montana, Wyoming, and Idaho.[7]

Their recovery is leading to what those individuals eager to have the grizzly delisted say are more frequent confrontations between people and grizzly bears, yet the reality suggests otherwise. As more people are using and recreating in grizzly bear recovery zones, the number of attacks hasn't changed much because more of us are becoming better educated about how to behave in grizzly bear habitat. Bear spray rental shops in YNP along with easy bear spray buying access at local stores encourages people to be prepared with bear spray and, while there are still many tourons and locals who behave badly, there are more visitors and residents alike willing to display fewer antagonizing behaviors when they see a bear. People are also being educated about the benefits of making noise and traveling in larger groups when they enter grizzly bear country.

As grizzly bears expand their range, the push to delist them for hunting suggests that the social carrying capacity for grizzly bears is still low. A reliance on nonlethal methods of protection against grizzly bears is on the upswing, but killing grizzlies because they have been habituated to human food, have killed livestock, threatened human lives or property, or were illegally poached has resulted in part in one-hundred-fifty-five dead grizzly bears between the years 2020 and 2022.[8] So many grizzly bears are killed each year without a hunting season that grizzly bear advocates see no reason to delist them. In the spring of 2022, two young grizzly bears jumped out of a dumpster that was not bear-proofed near Chico Lodge in Paradise Valley, north of YNP. The idea of grizzly bears at this commonly used dumpster site frightened people, so Montana's FWP planned to trap and relocate the bears. Grizzly bear advocates were so concerned someone would try to kill the bears that a volunteer with a gun stayed at the dumpster area overnight to protect the bears if they returned. The grizzly bears, fortunately, did not. Since these trash dumpsters in grizzly bear habitat were not bear-proofed, the bears could have potentially become habituated to human trash. Habituated bears tend to be dead bears.

A failure to increase the public's social carrying capacity for bears moving out of recovery zones makes population connectivity extremely challenging, just as it increases pressure on resource availability in established grizzly bear recovery zones. The result can be greater pressure on a zone's biological carrying capacity for grizzly bears, all while, due to increases in human-bear conflicts, decreasing the zone's social carrying capacity for the species. To Montana State Senator Willis Curdy, this reliance on relocating "problem" bears to existing recovery zones is exacerbating the situation by "stuffing one bear on top of another with a limited amount of resources for that bear to survive upon." [9]

For elk, a story rooted in tolerance defines their presence on Western lands. That story relied, in part, on nineteenth and early twentieth-century artists' depictions of a landscape that no longer existed, where the presence of elk as a metaphor for "wilderness, frontier, and lost nobility"[10] were being painted with increasing frequency all while elk numbers in the wild were rapidly disappearing. With more paintings of majestic bulls, rams, or bison, artists were able to build in the viewer's imagination a mirage. These painters perpetuated the essence of an American character and landscape that was tied to raw, wild, and open spaces full of wild animals to form what Andrew McKean, editor for *Outdoor Life,* called an "exaggerated reality." Like the nature fakers of Roosevelt's days, these artists, who McKean sees as natural history storytellers and hero-worshipers of a region, perpetuated the myth of the frontier as an integral part of the American spirit. To McKean, impressionist artists in the mid-nineteenth century relied on the imagery of a "muscular bull [elk] striking a heroic pose in a mythic landscape"[11] to capture the soul of the American West. Artists like Albert Bierstadt and Thomas Moran painted the West at the start of the Civil War as "Americans hoped it would be," while Carl Rungius, who, at the end of the nineteenth century, painted primarily for wealthy male sportsmen as a visual reminder of their experiences hunting there. Rungius's painting *Mountain Sheep,* completed in the early twentieth century and now on display in the Berkshire Museum in Massachusetts, portrays five bighorn rams in a pristine Western alpine environment. Bierstadt's paintings that depicted the existence of a reassuringly wild nature led him to become an important American landscape painter, while Rungius's paintings spanned the West's shift from wide-open wild landscapes to the arrival of ranches and towns. He relied on wildlife and wilderness as the centerpieces of his art, with his 1917 painting *The Days of Bison Millions,* on display at the National Museum of Wildlife Art in Jackson, Wyoming, depicting a large herd of bison that by 1917 were long absent from the plains. To the artists of the Hudson River School of Art that included Thomas Cole, Edward Moran, Carl Rungius, and Albert Bierstadt, their emphasis on creating nostalgic views of an untouched America before railroads and logging changed the landscape helped a broader audience encounter America's natural world. Their focus on wildlife and the beauty of American landscapes made them not only important figures in the developing conservation movement,[12] but by making elk essential to the West's mystique, assured them a place on a changing landscape (see figure C.1).

The practice of art that portrayed an artificial abundance of elk or bison offers another example of the shifting baseline syndrome, where in the case of scientists who define a species' reality based on their lived experiences rather than the species population in its natural state, these artists relied on past impressions to maintain another kind of illusion that offered wilderness scenes rich with very specific wildlife to convince consumers of their art that this version of the West remained intact. By constructing ideas about a species' rightful place, the losers were and continue to be many of America's wildlife species. Calling a coyote a "wilderness creature" creates the false idea that coyotes should not be living in our cities and towns even though for thousands of years they have benefited from following us around,[13] while

Figure C.1 Carl Rungius Painted *The Days of Bison Millions* in 1917 to Depict an Imagined Landscape of Days Long Past. Displayed at the National Museum of Wildlife Art, Jackson, Wyoming. *Source*: Barbara K. Jones.

labeling a cougar a mountain lion, according to Brian Silliman, an ecologist at Duke University, makes no sense either because "there is no such thing as a mountain lion." To him, the cougar was called a mountain lion because the mountains were the last place the animal lived. The language of "mountain lion" suggests they are mountain dwellers by choice, when they relied on other landscapes too, where they could find more prey.[14] By naming the cougar a mountain lion, a coyote a wilderness species, or a grizzly bear a high mountain animal assigns these animals to very specific ecosystems, while paintings that typically depicted elk in high country wilderness areas created an expectation of elk on those landscapes as well.

By relying on the concept of a species' native range, decisions about where wild animals can recover are based on ideas about where these species live today. What happens, though, if these animals were relegated to the Earth's margins because we gave them no other options? Hunting lore puts elk in the mountains when they historically occupied an extensive range from the mountains to the lowlands, while grizzlies are no longer considered to be plains animal, yet they thrived on bison while living there. Even in Canada, a plan for returning grizzlies to the prairie was never implemented after Environmental Canada's 2009 recovery strategy report identified human-caused mortality, habitat loss, and habitat degradation as the primary threats to prairie grizzly bear recovery. From the absence of habitat and sufficient natural food resources, the report concluded that the "*Recovery of this species is considered not technically or biologically feasible at this time.*"[15] This same logic plays out with grizzly bears and bison in Montana and Wyoming, where recovery and relocations are limited to socially accepted recovery or tolerance zones. A reliance on contemporary range maps or tolerance zones for species in recovery and the language of native versus non-native creates additional human-defined boundaries that challenge an animal's survivability in a rapidly changing world.

This idea about where species are supposed to live is closely tied to our expectations about wildlife's place. Caribou, a species that no longer lives in the lower forty-eight states, has been part of America's natural history since the early Holocene. Research done in the 2000s by University of Michigan archaeologist John O'Shea identified caribou migrations that would have traveled across the Upper Peninsula of Michigan into what is now Lake Huron. By relying on side-scan sonar and divers who explored the Alpena-Amberley Ridge in the lake, a series of what appeared to be human-made stone structures were identified. These structures created a bottleneck for migrating caribou on a once-dry ridge that formed what O'Shea called a "funnel lane." This lane directed caribou into corrals and past blinds, so hunters were invisible to them.[16] Ten-thousand years ago, caribou had a place in Michigan, but by 1912, through habitat destruction, hunting, and vehicle collisions, boreal woodland caribou were pushed north onto islands in Ontario's region of Lake Superior. On a few of those northern islands there existed a robust island population of about one thousand caribou that today, through the forces of predation and climate change, is down to about sixty. The American and Canadian governments have relied on a "wait and see" approach to caribou that has done little to slow the decline in caribou numbers.

According to the Biigtigong Nishnaabeg and Michipicoten First Nations, the caribou are adapted to older, conifer-dominated forests, but due to human disturbance, these older forests are often being replaced, through logging[17] and development, with younger deciduous or mixed deciduous-conifer forest areas. A "wait and see" policy has allowed these younger forests to expand and in turn encouraged the range expansion of moose and white-tailed deer at the expense of caribou. By default, the caribou's place has become ill-suited for its own survival. For the Biigtigong Nishnaabeg and Michipicoten First Nations, a willingness of the Canadian government to shift away from a wait and see approach could potentially halt the caribou's southern range recession in Canada and demonstrate that caribou recovery is possible. With that, the likelihood of this population's extirpation lessens.

THE METAPHORS OF PLACE

The metaphors and stories we share are powerful forces that serve to maintain our understanding of the world and reaffirm who and what belong in specific places. Those in California who don't like or want beavers around can rely on the language of "non-native" to remove beavers that are outcompeting "native" species for resources, while in Oregon a small number of trappers have been successful at arguing for trapping as their way of life so they can continue to trap beavers on federal land. For elk in Montana, an idea that elk are a mountain forest species that has been pushed to the lowlands by wolves is one argument hunters have used to kill high numbers of wolves, while in Wyoming wolves are trophies in the northwestern part of the state and predators elsewhere. With generous state laws regarding wolf hunting, the language of trophy and predator further encourages their removal. For Wyoming hunters, even though many hunting groups want to change the stigma associated with shooting wildlife as trophies, black bears, grizzly bears, mountain lions, and wolves are all designated as trophies, with grizzly bears a current exception[18] because they are still protected under the ESA.

The language of management also looks at wildlife as a consumable product. When wildlife killing through hunting is called "harvesting" or pregnant does in Wyoming that are dying because of severe weather means "losing an entire fawn crop,"[19] agencies and state officials have shifted hunting away from killing wild animals to harvesting a crop. Although harvesting crops might be a way to see hunting on anthropogenic landscapes, it is not a useful way of seeing wild nature and wilderness. It reminds us that the continued reliance on a system of wildlife management rooted in agriculture and forestry and the principles of producing and harvesting for the greatest sustainable yield, while it does sanitize hunting as a lifestyle choice for many, is not the best language for encouraging conservation or coexistence. Anti-predator paranoia also does not encourage better management practices; instead, it allows state legislators to focus on how many wolves, mountain lions, and bears can be killed rather than how many can live.[20] In January 30, 2025 at a Montana House Fish, Parks, and Wildlife committee meeting, one House member proposed "An act revising laws related to the mangement of gray wolves; requiring the Fish and Wildlife Commission to apply different techniques for the management of gray wolves; allowing the use of thermal and infrared scopes for gray wolf management" (HB 259). This change would allow military-grade technology to be used so wolves could more effectively be killed at night. For the proposing legislator, killing wolves wasn't "about fair chase, it's about [giving] management" more tools to reduce wolf numbers. He went on to add that like with "rats or other pests, it is about managing to control a problem." By not relying on scientific evidence, Montana officials have never demonstrated that the state is being overrun by wolves. This lack of evidence has forced many people to question what problem the state is trying to "control" by this change. In addition, by admitting HB 259 is not about fair chase, it becomes appparent that wolf management is not constrained by hunting ethics. As I am writing this, the proposed bill passed the Montana House but not yet the Senate. At a House Floor Session the next day for HB 176 called "Require unlimited wolf hunting quota when population is at or above 450 wolves," even though one House member said the bill "looks gross to me" and another said this bill "sounds like a hunt, baby, hunt bill," the bill moved forward in the House with a vote of 57 to 42. Such political strategies do not align with the hunting is conservation metaphor or any effort at increasing the social carrying capacity for predators.

To Jason Baldes, an Eastern Shoshone and executive director for the Wind River Tribal Buffalo Initiative, Western wildlife management strategies rely on a silo approach that breaks ecosystems into separate parts. Baldes doesn't see nature as siloed, but as one system working together. By removing predators that can include bears and wolves, as well as bobcats and cougars, from an ecosystem to manage for elk or deer, the predator's role in weeding out the old, weak, or sick, or keeping ungulates moving so they live across the entirety of an ecosystem is no longer maintained, while locking bison behind fences keeps them from fulfilling their ecological role as a keystone species for prairie ecosystems. Engineered resilience through silo-style management practices looks at the parts of the ecosystem as separate entities, while natural resilience maintains a healthy ecosystem by recognizing how the parts work together to form one complete ecosystem. To Christine Peterson, writing in *High Country News* about the role predators have in containing deadly animal disease, "Only a complete ecological system can ensure that all species have a chance against threats like disease and even climate change."[21] The human paranoia associated with predators or an intolerance for wild bison, beaver, or caribou strips those species of their ecosystem membership and in turn weakens the resilience of the landscapes that have relied on those species to make them whole.

NEW STORIES, NEW METAPHORS

If we can move toward Wendell Berry's "contrary assumption"[22] that what is good for the world will be good for us, we will have made a major step toward seeing wildlife and nature through a more mutualistic lens. We can even stretch that to be a form of coevolved mutualism, where we live in a world of interdependent relationships that exist between us and nature. Since so much of the world today is managed by an ever-growing human population, it can be challenging to imagine that not only are we a part of nature, but we are dependent on it. Nature and wildlife provide us with everything from clean air and water to manatees in Florida's springs and caribou in our boreal woodlands, yet by denying them the space they need to survive, we reduce their ability to function and in turn our own well-being. Black bears returning to Big Bend National Park on their own because they have a place to return to is an example of us giving black bears both space and opportunity. To Tom VandenBerg, chief of Big Bend's interpretation and visitor services, "[Their return] is really one of the great feathers in the cap of Big Bend National Park."[23] Since black bears or woodland caribou are charismatic flagship species of their ecosystems, by sharing spaces with them, we are reciprocating nature's generosity to us.

In a book called *About Looking* when the author, John Berger, wrote, "Animals first entered [our] imaginations as messengers and promises," he was writing about symbolic thought, where animals were our first symbols.[24] To the environmental journalist Ben Goldfarb, we relied on them for food, but they were also our metaphors, symbols, and partners. They were built into our myths, stories, and relationships because animals sustained us biologically and culturally. Yet a reliance on dominion over nature as a metaphor for seeing the world has allowed us to kill animals in unsustainable ways and, with that loss, become disconnected from them and all they provide. What started with us and nature as members in a shared club turned into an exclusive us-only club so that today, even though we have the Animal and Plant Health Inspection Service (APHIS) or the Interagency Bison Management Plan to manage "problem" wolves or wild bison, wildlife managers are still expected to make additional rules regarding the movement and survival of predators and perceived "pests." If we define paranoia to be mistrust, hypervigilance, and difficulty in forgiveness, those who are unwilling to coexist with wildlife need new stories that replace mistrust with trust, hypervigilance with acceptance, and difficulty in forgiveness with a willingness to let go of past slights that forged a cultural heritage ill-suited to today's world.

The metaphors wolf as vermin or bison as livestock are examples of ways of seeing the world rooted in the agrarian ideal, where hard work and production are tied to moral virtue. When Thomas Jefferson argued that the yeoman farmer as the primary occupant of the land in America was best suited for balancing the farmer's self-interest with the common good, he was referring to, like Wendell Berry, the social ties and good land stewardship of rural America that relied on honesty, integrity, and charity to survive. The persistence of the agrarian ideal metaphor in the face of an eroding regard for the common good, as well as a decline in transparency in wildlife management decisions, speaks to its long-lasting power. With each of the metaphors of ranching as a way of life, hunting as conservation, and cattle is king, the public good is subsumed under ways of seeing the world rooted in self-interest. The reliance on the metaphor cattle is king keeps a non-native animal on public land at the expense of bison, while the metaphor ranching as a way of life or hunting as conservation can rely on anti-predator traditions and cultural heritage that tie manliness and productivity to morality and dominance over nature to maintain their legitimacy.

At a Montana FWP Wolf Management Plan Scoping Meeting I attended, of the thirteen people who spoke, eleven of them wanted FWP to create a plan that prioritized live wolves on Montana's land rather than continuing to rely on lethal management strategies. The speakers highlighted how wolves are not managed as public trust resources for all Americans when on federal lands, how the lack of transparency from FWP about wolf numbers makes an accurate wolf count unavailable to most wolf advocates, particularly when, in many cases, the numbers are determined by hunters' observations, and how FWP's wolf education is geared toward hunters and trappers rather than to educating all Montanans about the benefits of wolves on the land. To a speaker from Trap-Free Montana, "If [the FWP] doesn't advocate for wolves, who will?" This lack of advocacy has also allowed inaccurate stories about declining elk numbers because of wolves to become a powerful message hunters rely upon, but without pushback from FWP those narratives have become sticky and hard to shift. When two speakers in the meeting argued that the decline in elk numbers was the result of too many wolves, a wolf advocate responded with "that is nonsense when you know that five elk management areas exceed elk population objectives and two are close to it." To this speaker, the "FWP should call them out" for promoting "misinformation." For wolf advocates, the FWP's unwillingness to correct inaccurate species' information indicates a bias driven by politics rather than science, particularly when, according to the FWP's own 2021 elk count, eleven of the state's elk survey units were at more than double their population targets.[25]

To encourage greater wolf tolerance and efforts at coexistence, a shared idea across all wolf advocacy speakers was the creation of a wolf advisory group made up of stakeholders whose decisions regarding wolf management would be considered in any new management plan. A speaker for Wolves of the Rockies suggested maintaining a stakeholder balance where every consumptive user of the wolf resource would be offset by a non-consumptive user, while the Trap-Free Montana speaker felt that approach gave hunters and trappers too much of a voice when most Montanans don't even hunt or trap wolves. Advocates for the wolves were not limited to Wolves of the Rockies, Trap-Free Montana, or the Bear Creek Council, but included ecotour operators and guides who believe the benefits to Montana's economy from ecotourism receipts need to be considered when wolf management decisions are being made. A speaker for Wild Livelihoods expressed concerns that the economic impacts of tourism are typically not included in wildlife management plans, while a YNP guide pointed out how livestock receipts are counted in wildlife management decision-making but tourist receipts that contribute significantly to Montana's economy are not. The guide's belief that tourist receipts are ignored in wildlife management decisions is indicative of the power of livestock and the ranching metaphor. To wolf advocates, by allowing the claim of "too many wolves" and "too few elk" to stand and by not determining the actual carrying capacity for wolves in Montana, the FWP has prioritized the voices of hunters, trappers, and ranchers over that of educating Montanans about the value of coexistence and the many benefits wolves provide.

Curiously, even after this discussion, one wolf advocate felt "there was quite a bit of tolerance for wolves [in Montana] even though current plans [and laws] suggest otherwise." The results from a January 11, 2024 survey through Montana's FWP and the University of Montana's College of Forestry and Conservation support this individual's anecdotal evidence. Based on the results, 74 percent of those surveyed (including deer and elk hunters, landowners, wolf hunters and trappers, and other Montanans) were tolerant of or very tolerant of wolves in Montana. Yet when it comes to state agencies and elected officials, it is apparent the attitudes of the Old West still prevail over those of the New.

Something similar is continuing to happen with bison that leave YNP. The bison-as-livestock metaphor helps explain why YNP's park superintendent, Cam Sholly, said in a *New York Times* article about the winter of 2022–2023 bison hunt, "[It] is the only species we constrain to a boundary." The severe winter weather of 2022–2023 pushed the bison north across the boundary out of YNP along their traditional migratory path into Beattie Gulch, an area adjacent to Gardiner, the Park's northern gateway town. The large numbers of bison seeking food forced the animals to spread out across the Gardiner basin, where there are bison tolerance zones. Beattie Gulch is a tolerance zone, as are others that are very difficult for bison to access. In the past, to prevent bison from getting near livestock on ranches in Paradise Valley, Montana, the animals that left the park were hazed back into the park or sent to slaughterhouses. With today's increased hunting opportunities for treaty tribes that are allowed to hunt on unclaimed treaty lands along with the transfer of bison to tribal lands, the need for that kind of slaughter has lessened. Although the park wants to get out of the business of slaughtering bison at slaughterhouses, the Confederated Salish and Kootenai Tribes do not want the park's slaughter program to end because they rely on some of that bison meat for their school lunch programs (see figure C.2).

At YNP's northern entrance, bison can walk through the gates and head over the bridge into the town of Gardiner, or they can migrate past the local

Figure C.2 Yellowstone National Park's Boundary Adjacent to the Town of Gardiner, Montana. *Source*: Barbara K. Jones.

school and end up in Beattie Gulch. To those opposed to the hunt, another kind of slaughter occurs when bison move into the "killing fields" of Beattie Gulch. Mike Mease of the Buffalo Field Campaign, an organization working to stop the slaughter and harassment of Yellowstone's bison herds and protect the natural habitat of wild, free-roaming bison, recognizes the importance of the tribal hunt but criticized what he calls the "commercial influence" driving the hunt. To him, "No other wildlife is treated this way. This is all directed by the Montana livestock industry."[26]Critics of the hunt argue that the limits placed on the bison's migration routes outside the park trap the bison in a small space for a public and tribal hunt that defies the rules of fair chase. In support of the wild YNP bison, some landowners in Beattie Gulch have posted orange signs that read "Buffalo Safe Zone, No Hunting or Harrassing of Bison By Any Person or Agency." The signs include a drawing of two bison and a calf with the words: "By Order of the Governor of Montana: The Department of Livestock May Not Enter Private Property Without Landowner Permission Unless There is an Immediate Threat of Disease Transmission from Bison to Livestock or People." These posted "safe zones" are in place to offer bison shelter from being killed or otherwise bothered while in Beattie Gulch.

For treaty tribe members, who represent most of the bison hunters, the hunt allows them to not only reconnect with their families and restore their relationship with bison but to also fill their freezers. However, it also creates unhealthy competition between treaty tribes as they work to maximize bison hunt success. Since there is no reward for killing fewer bison than necessary and no shared goals between the treaty tribes, the hunt can become another tragedy of a common resource. To some bison hunt officials from YNP and the Greater Yellowstone Coalition (GYC), the inability of treaty tribes to self-regulate or develop a shared management plan for the bison hunt allowed the killing of bison to exceed twenty-five percent of the total number that both the park and the GYC felt was enough. The large numbers of bison killed led the park to "hold back" or deny hunting access to hundreds of additional bison.

The need to slaughter, hunt, or transfer brucellosis-free bison to tribes is the result of the low social tolerance that exists for bison from politicians and many ranchers across Montana. For those trying to better manage bison that leave the park, the feeling is "the state is not a willing partner." According to Shana Drimal, Montana Governor Gianforte's opposition to bison "supersedes any of the other wildlife issues in Montana" to the extent that she believes he is willing to "fall on the sword for wolves, but not bison." Since no brucellosis outbreak in Montana's two- three million cattle has ever been traced back to Yellowstone bison, to Sholly, it is curious that preventative measures imposed on bison against brucellosis have typically not been considered when it comes to the elk's role in transmission. In that same earlier *New York Times* interview, Sholly added, "It's hard to claim bison are presenting an imminent threat to livestock while thousands of brucellosis-infected elk are literally side-by-side with livestock in the Paradise Valley and there is no strategy to manage that interface." Shifting the metaphor bison as livestock to one that sees bison as a wild animal will take a significant increase in social tolerance from both politicans and ranchers. (see figure C.3).[27]

For The Nature Conservancy, their ownership of the 60,000-acre Matador Ranch in the American Prairie restoration area has offered local ranchers a way to see coexistence with bison and other wildlife in a more positive light. By creating a "grassbank," ranchers may graze their cattle on the Conservancy's land at a low cost in exchange for a pledge to follow wildlife-friendly management practices on their own land. The collective model of large-scale prairie protection initiated by American Prairie and expanded upon by groups like The Nature Conservancy have formed new opportunities for ranchers and encouraged them to use wildlife conservation strategies on their own ranches. By working together these wildlife-friendly practices have spread to an additional 240,000 acres of private property within American Prairie's recovery

Figure C.3 In a Free Speech Zone in Yellowstone National Park, the Buffalo Field Campaign Has a booth to Educate People about Bison. *Source*: Barbara K. Jones.

area.[28] As communities build connections like these that are rooted in coexistence, they can forge a "New West" cultural heritage worth protecting.

By making coexistence and a greater sense of mutualism our default relationship with wildlife and nature, new stories will emerge that can help maintain these relationships. Woodland caribou as gray ghosts made them invisible to us, beavers as pests discounted their skills as ecosystem engineers, while prairie dogs as varmints highlighted their unpopular behaviors rather than their talents at enhancing grass quality and building habitats for other species. However, new, more natural discourse can help shift these relationships. They can overwhelm old stories and myths, changing how we see these animals and their ways of living so the outdated figurative language of the past becomes less powerful and less persistent. Unnatural discourse has been used to maintain and encourage a specific worldview that keeps people separate from nature on anthropogenic landscapes; yet by accepting that we are a part of nature, the interdependence of people and the natural world will make coexistence stories the ones we are more likely to tell.

As a final and somewhat reassuring word, it is important to mention that in 2023 on International Beaver Day, a beaver named Barbie was named the year's climate hero, and Bubbles is serving as a coexistence consultant. Since language and names matter, more than thirty years after a beaver pair was named Adolf and Eva by Beaver Scouts at the Gateshead Garden Festival in England, it is now up to Barbie and Bubbles to help us reimagine not only beavers, but every other maligned species as critical partners in our shared natural world.

NOTES

1. WGFD News, April 15, 2024. https://wgfd.wyo.gov/news-events/grizzly-bear-euthanized-due-cattle-depredation-washakie-county.
2. Eggert, Feb 10, 2023.
3. Auster et al., p. 3.
4. Burnham and Mott. April 2, 2021 (updated Nov 2, 2021).

> "Federal scientists who track grizzly bears in the Greater Yellowstone Ecosystem are revising a key criterion that underlies how they count bears, with the effect being a 34% to 43% jump in the population estimate. The change brings the estimated population from 727 grizzlies up to more than 1,000 animals. The protocol used to count grizzlies—called the Chao 2 method—is staying the same, continuing a method that's been used since 2007. What's changing is a filter that determines how many female grizzlies with cubs get plugged into that model. Historically, sows and cubs spotted from the air within 19 miles of each other were only tallied once to avoid double-counting bears. But an Interagency Grizzly Bear Study Team analysis found that this buffer was too broad. So going forward only sows with cubs detected within 10 miles of each other will be excluded to avoid duplication." Koshmrl. April 12, 2021

5. Rovelli. C5.
6. Eggert. December 12, 2022.
7. 2022 Draft MT-FWP Grizzly Bear Management Plan. 40–41.
8. Grizzly & Wolf Discovery Center, West Yellowstone, MT, June 2023–.
9. Eggert. April 15, 2021.
10. McKean 59, 60.
11. McKean 1.
12. McKean 62.
13. Flores 8.
14. Pagnamenta 167.
15. Recovery Strategy for the Grizzly Bear (Ursus arctos), Prairie Population, in Canada. iii.
16. Brackley 30.
17. Logging in Canada is destroying the boreal forests that act like carbon sinks, all while adding 10 percent of the country's greenhouse emissions. The logging industry and the Canadian government promoted the myth that clearcutting is sustainable, even though the consequences for Indigenous communities and wildlife habitat are dire (NRDC Spring 2023).
18. There are exceptions to this, most recently when a shot grizzly bear was found on the North Fork Highway on May 2, 2023, near Cody, Wyoming. Public outcry has been intense over this poaching incident.
19. Eggert. April 12, 2021.
20. Eggert. February 10, 2023.
21. Peterson 40.
22. Berry 2002.
23. Brulliard 18.

24. Berger 2, 7.
25. Eggert, September 8, 2022.
26. Robbins *New York Times*. April 4, 2023.
27. Robbins *New York Times*. April 4, 2023.
28. Jachowski, *The Conversation*. April 26, 2023.

Appendix

List of Interviews

Lydia Allen, Caribou Activist
Chamois Andersen, Defenders of Wildlife's Rockies and Plains Senior Representative
Malou Anderson-Ramirez, Montana Rancher and Mother
David Argument, Resource Conservation Manager at Parks Canada
Dr. Roger Auster, Lecturer in Environmental Social Science, University of Exeter
Mike Callahan, Founder of the Beaver Institute, Inc
Cory DeStein, a director for the Arrow Lakes Caribou Society in British Columbia, Canada and a professional photographer
Shana Drimal, Senior Wildlife Conservation Associate at the Greater Yellowstone Coalition
Greg Finstad, University of Alaska Fairbanks professor and Reindeer Research Program Manager
Donald Forgione, Paynes Prairie State Park Manager
Dr. Suzanne Fouty, Retired Forest Service Hydrologist/Soils Specialist
Noah Greenwald, Endangered Species Director at the Center for Biological Diversity
Aaron Hall, Beaver Specialist for Defenders of Wildlife
Mark Heinz, Wildlife and Outdoors Reporter for Wyoming's *Cowboy State Daily*
Rick Lamplugh, Author and Bison Activist
Matthew La Roche, Retired Superintendent of the Allagash Wilderness Waterway, and a Registered Maine Guide
Mark McCollough, Retired Endangered Species Biologist. Maine Field Office, U.S. Fish and Wildlife Service
John Pastor, Retired Minnesota State Wildlife Biologist
Mike Schrage, Wildlife Biologist for the Fond du Lac (Ojibway) Reservation Management Division
Gregg Servheen, Retired Wildlife Biologist and Natural Resource Manager for the Idaho Department of Fish and Game
Mark Vieira, Carnivore and Furbearer Program Manager for Colorado Parks and Wildlife

Bibliography

Abrams, Jesse, Michael Johnduff, and Susan Charnley. *Beaver-Related Restoration in Owyee County*. Idaho: Opportunities and Challenges. USDA Forest Service. Pacific Northwest Research Station. Research Paper PNW-RP-611. February 2019.

Arendt, Hannah. Reveal Meaning without Defining it. *Nico*. November 26, 2015. https://universalstoryteller.com/2015/11/26/reveal-meaning-without-defining-it/. (Accessed October 27, 2021).

Arnold, Billy. Y'stone Grizzly Attacks are Still Rare. *Jackson Hole News & Guide*. August 16, 2023. 1A, 10A-11A.

Ashby, Jack. *Platypus Matters: The Extraordinary Story of Australian Mammals*. Chicago: The University of Chicago Press. 2022

Auster, Roger E., Stewart Barr, and Richard E. Brazier. Improving Engagement in Managing Reintroduction Conflicts: Learning from Beaver Reintroduction. *Journal of Environmental Planning and Management*. Vol 64 (10). 2021. 1713–1734

——. Renewed Coexistence: Learning from Steering Group Stakeholders on a Beaver Reintroduction Project in England. *European Journal of Wildlife Research* 68 (1). December 3, 2021.

Auster, Roger E., et al. Learning to Live with Reintroduced Species: Beaver Management Groups are an Adaptive Process. *Restoration Ecology: The Journal of the Society for Ecological Restoration*. Vol 31 (5): 1–6. March 10, 2023.

Backhouse, Frances. *Once They Were Hats: In Search of the Mighty Beaver*. Toronto: ECW Press. 2015.

Bachman, Melissa. Prairie Dog Plinking – Great Summertime Rifle Practice. https://winchester.com/Blog/2017/07/prairie-dog-plinking-great-summer-rifle-practice. (Accessed August 25, 2022).

Bailey, James A. *American Plains Bison: Rewilding an Icon*. Helena: Farcountry. 2013

BARK. Beaver are Climate Allies but are Treated as Pests in Oregon. https://bark-out.org/get-involved/beaver-are-climate-allies-not-pests/. (Accessed June 5, 2022).

Bartram, William. *Travels of William Bartram.* Mark Van Doren (ed). New York: Dover Publications. 1955.

Beaver Institute. https://www.beaverinstitute.org/. (Accessed June 20, 2022).

Berry, Wendell. *The Unsettling of America: Culture and Agriculture.* San Francisco: Sierra Club Books. 1996.

——. *The Art of the Commonplace: The Agrarian Essays of Wendell Berry.* Berkeley, CA: Counterpoint. 2002.

Blair, Neal L. *The History of Wildlife Management in Wyoming.* Cheyenne: Wyoming Game and Fish Department. 1987.

Booher, Charlie. Hunting is Conservation, but Is It Enough? *The Bugle.* January 2022. 37–39.

Boyd-Fleigel, Sophia. New Laws. *Jackson Hole News & Guide.* March 8, 2023. 8A–9A.

Brackley, Chris, and Jake O'Flaherty. Life Before Lake Huron. *Canadian Geographic.* March/April 2023. 30–31.

Braitman, Laurel. *Animal Madness: How Anxious Dogs, Compulsive Parrots, and Elephants in Recovery Help Us Understand Ourselves.* New York: Simon & Schuster. 2014.

Braverman, Irus. *Wild Life: The Institution of Nature.* Stanford: Stanford University Press. 2015.

Breining, Greg. *Return of the Eagle: How America Saved Its Natural Symbol.* Guilford, CT: Falcon Publishing, 1994.

Brewster, R. Kyle, et al. Do you hear what I hear? Human perception of coyote group size. *Human–Wildlife Interactions.* 11(2):167–174, Fall 2017.

Brinkley, Douglas. *The Quiet World: Saving Alaska's Wilderness Kingdom 1879-1960.* New York: Harper Collins, 2011.

Bromley Horn Dance. https://tunearch.org/wiki/Annotation:Abbot%27s_Bromley_Horn_DanceThe Abbots Bromley Horn Dancers. (Accessed December 15, 2022).

Browne, William P., et al. *Sacred Cows and Hot Potatoes: Agrarian Myths in Agricultural Policy.* San Francisco: Westview Press. 1992.

Brulliard, Nicolas. Comeback Bears. *National Parks.* Spring 2023. 18–19.

Burnham, Josh, and Nick Mott. Timeline: A History of Grizzly Bear Recovery In The Lower 48 States. *Montana Public Radio.* April 2, 2021 (updated November 2, 2021). https://www.mtpr.org/montana-news/2021-04-02/timeline-a-history-of-grizzly-bear-recovery-in-the-lower-48-states. (Accessed March 13, 2023).

Byrd III, Colonel William. As Quoted in Mitsch, William J. and James G. Gosselink. *Wetlands.* New York: Van Nostrand Reinhold. 1993. 12.

Callahan, Mike. A New Idea for Beaver Coexistence – Beaver BioBanks! *The Beaver Institute Blog.* https://www.beaverinstitute.org/a-new-idea-for-beaver-coexistence-beaver-biobanks/. (Accessed September 1, 2022).

Campbell, R., A. Dutton, and J. Hughes. *Beavers and Tourism in Economic Impacts of the beaver.* Oxford: University of Oxford. 2007.

Capuzzo, Michael. *Close to Shore: A True Story in an Age of Innocence.* New York: Broadway Books. 2001.

Cecco, Leyland. Polar Bear Numbers in Canadian Arctic Pose Threat to Inuit, Controversial Report Says. November 18, 2013. https://www.theguardian.com/

world/2018/nov/13/polar-bear-numbers-canadian-arctic-inuit-controversial-report. (Accessed April 15, 2022).

Chaline, Eric. *Fifty Animals that Changed the Course of Human History*. Buffalo: Firefly Books. 2011.

Chaney, Robert. *The Grizzly in the Driveway: The Return of Bears to the Crowded American West*. Seattle: University of Washington Press. 2020.

Chivers, C. J. Fear on Cape Cod as Sharks Hunt Again. *New York Times Magazine*. October 20, 2021.

Clark, Susan G. Saving Ourselves by Saving Nature in Headwaters. *Jackson Hole News & Guide Conservation in Greater Yellowstone Special Issue*. 2021. 6–7.

Cochrane, Willard W. *The Development of American Agriculture: A Historical Analysis*. Second Edition. Minneapolis: University of Minnesota Press. 1993.

Coleman, Jon T. *Vicious: Wolves and Men in America*. New Haven: Yale University Press. 2004.

Collins, Dac. Montana's Controversial Pheasant Stocking Program Moves Forward. *Outdoor Life*. July 23, 2022. https://www.outdoorlife.com/conservation/montana-pheasant-stocking-program-moves-forward/#:~:text=Montana's%20Controversial%20Pheasant%20Stocking%20Program%20Moves%20Forward,-The%20state's%20plan&text=Montana%20Fish%20Wildlife%20and%20Parks,effort%20to%20recruit%20more%20hunters. (Accessed August 2, 2022).

Coyote Cowardice. *Jackson Hole News & Guide*. March 8, 2023. 5A.

Cronon, William. Revisiting the Vanishing Frontier: The Legacy of Frederick Jackson Turner. *Western Historical Quarterly*. Vol. 18, No. 2. April 1987. 157–176.

Davis, Joseph. Covering Sharks in Shark Week's Wake. *Society for Environmental Journalists*. July 27, 2022. https://www.sej.org/publications/tipsheet/shark-week-you-are-going-need-bigger-meme. (Accessed August 5, 2022).

Defenders of Wildlife. Bison. https://defenders.org/wildlife/bison#:~:text=Though%20bison%20once%20roamed%20across,and%20Fort%20Belknap%20Indian%20Reservations. (Accessed March 9, 2023).

Dekret, Vaughn. Gifford Pinchot. https://dnr.maryland.gov/centennial/pages/centennial-notes/giffordpinchot.aspx. (Accessed April 7, 2022).

Devereaux, Ryan. Was Yellowstone's Deadliest Wolf Hunt in 100 Years an Inside Job? *High Country News*. July 27, 2022. https://www.hcn.org/articles/wolves-was-yellowstones-deadliest-wolf-hunt-in-100-years-an-inside-job.

——. Bear Minimum. *The Intercept*. https://theintercept.com/2023/01/29/grizzly-bear-endangered-montana/. (Accessed February 1, 2023).

Dickson, Tom. Leave it to Beavers. *Montana Outdoors*. May–June 2022. 19–23.

Diving Blog. *5 Things More Likely than a Shark Attack*. http://thedivingblog.com/5-things-more-likely-than-shark attack/. (Accessed September 9, 2016).

Downey, Hannah. Public Comment of American Prairie Preserve Bison Grazing. *PERC*. July 22, 2021. https://perc.org. (Accessed August 13, 2022).

Dupzyk, Kevin. Business is Blooming. *Audubon*. Spring 2023. 34–41.

Dryzek, John S. *The Politics of the Earth: Environmental Discourses*. Oxford: Oxford University Press. 1997.

Earth Justice. How We Helped Bison Make a Huge Comeback. October 5, 2021. https://earthjustice.org/features/bison-recovery-prairie#:~:text=In%202012%2C%20after%20years%20of,Fort%20Belknap%20Reservations%20in%20northern. (Accessed November 20, 2021).

Eggert, Amanda. Snares Legalized and Trapping Season Expanded. *Montana Free Press*. April 12, 2021. https://montanafreepress.org/2021/04/12/snares-legalized-and-trapping-season-expanded/. (Accessed May 12, 2023).

——. Legislature Passes Black Bear, Grizzly Bear, and Wolf Proposals. *Montana Free Press*. April 15, 2021. https://montanafreepress.org/2021/04/15/legislature-passes-black-gear-grizzly-bear-and-wolf-proposals/. (Accessed May 12, 2023).

——. BLM Approves American Prairie Reserve's Bison Grazing Proposal. *Montana Free Press*. July 29, 2022. https://montanafreepress.org/2022/07/29/blm-approves-american-prairie-reserve-grazing-application/. (Accessed August 3, 2022).

——. Elk Management in the Crosshairs. *Montana Free Press*. September 8, 2022. https://montanafreepress.org/2022/09/08/elk-management-fwp-spotlight/. (Accessed September 10, 2022).

——. Wildlife Watch: Grizzly Bear Management. *Montana Free Press*. December 12, 2022. https://montanafreepress.org/2022/12/12/wildlife-watch-grizzly-bear-management/. (Accessed January 20, 2023).

——. The Grizzly Rorschach Test. *Montana Free Press*. February 10, 2023. https://montanafreepress.org/2023/02/10/montana-grizzly-bear-management-plan/. (Accessed March 1, 2023).

Einhorn, Catrin. A Nevada Rancher Made a Truce with Beavers, and It Paid Off. *New York Times*. September 6, 2022. A1, A15.

England, Dan. When a Pet Becomes a Problem: Nuisance Species, Many of Them Former Pets, Wreak Havoc on Wyoming's Wildlife. *Wyoming Wildlife*. July 2022. 26–31.

Environmental Assessment. American Prairie Reserve Change of Use DOI-BLM-L010-2018-007-EA. June 2021. Section 2.6.

Federal Register. Endangered and Threatened Wildlife and Plants; Determination of Endangered Status for the Population of Woodland Caribou Found in Washington, Idaho, and Southern British Columbia. Vol. 48, No. 206. October 25, 1983. 49245–49247.

First Nations and David Suzuki Foundation. *The Cultural and Ecological Value of Boreal Woodland Caribou Habitat*. July 2013. https://davidsuzuki.org/wp-content/uploads/2013/07/cultural-ecological-value-boreal-woodland-caribou-habitat.pdf. (Accessed March 12, 2023).

Florida Wildlife Corridor Foundation. https://floridawildlifecorridor.org/. (Accessed September 2, 2022).

Frodeman, Robert. Guest Shot. *Jackson Hole News & Guide*. June 16, 2021.

Fountain, Steven M. Ranchers' Friend and Farmers' Foe: Reshaping Nature with Beaver Reintroduction in California. *Environmental History*. Vol. 19, No. 2. April 2014. 239–269.

Gagliardi, Robert. History of Wildlife Management. *Wyoming Wildlife*. April 2022. 14–15.

Gocke, Mark, and Doug McWhirter. Completing the Circle. *Wyoming Wildlife*. May 2022. 38–44.

Goldfarb, Ben. *Eager: The Surprising, Secret Life of Beavers and Why They Matter*. White River Junction. NY: Chelsea Green Publishing. 2018.

——. Can Camera Traps Relieve Our Species' Loneliness? *High Country News*. March 2023. 19–21.

Goldfarb, Ben and Biographic. Re-beavering the American West. *The Atlantic*. December 4, 2018. https://www.theatlantic.com/author/ben-goldfarb/. (Accessed November 21, 2022).

Gómez-Upegui, Salomé. Florida's Manatees are Dying in Record Numbers – but a Lawsuit Offers Hope. *The Guardian*. June 10, 2022. https://www.theguardian.com/environment/2022/jun/10/floridas-manatees-are-dying-in-record-numbers-but-a-lawsuit-offers-hope#:~:text=Florida's%20manatees%20are%20dying%20in%20record%20numbers%20%E2%80%93%20but%20a%20lawsuit%20offers%20hope,-This%20article%20is&text=The%20US%20Fish%20and%20Wildlife,to%20die%20in%20record%20numbers. (Accessed August 15, 2022).

Gow, Derek. *Bringing Back the Beaver: The Story of One Man's Quest to Rewild Britain's Waterways*. White River Junction, VT: Chelsea Green Publishing. 2020.

Greenaway, Guy, Greg Chernoff, and Ken Sanderson. Representing Ecosystem Services: Three Perspectives on Modeling the Benefits of Nature. *Miistakis Institute*. June 2013.

Greenberg, Joel. *A Feathered River Across the Sky: The Passenger Pigeon's Flight to Extinction*. New York: Bloomsbury, USA. 2014.

Grizzly & Wolf Discovery Center. West Yellowstone, MT June 2023.

Grug, Mari. Licensed Beavers Released in Wales for the First Time. March 30, 2021. https://www.bbc.com/news/uk-wales-56565050. (Accessed May 29, 2022).

Grunig, James E. Situational Theory on Publics. January 2005. file:///C:/Users/docjo/Downloads/SituationalTheoryofPublicsfromEncyclopediaofPR.pdf. (Accessed May 20, 2021).

Grunig, James E., and Keith R. Stamm. Cognitive Strategies and the Resolution of Environmental Issues: A Second Study in *Journalism and Mass Communication Quarterly*. December 1979. file:///C:/Users/docjo/Downloads/CognitiveStrategiesandtheResolutionofEnvironmentalIssuesASecondStudyJMCQ1979.pdf. (Accessed April 25, 2021).

Haddock, Rachelle. Beaver Restoration Across Boundaries. Miistakis Institute. March 2015. https://www.beaverinstitute.org/wp-content/uploads/2021/03/Haddock-2015-Miistakis-Institute-Beaver-Restoration-Across-Boundaries.pdf. (Accessed May 22, 2022).

Hall, Travis. Can Prison-farmed Pheasants Save Hunting? *Montana Free Press*. June 16, 2022. https://montanafreepress.org/2022/06/16/montana-fwp-fields-opposition-to-pheasant-stocking-operation/. (Accessed August 15, 2022).

Hanna, Warren Leonard. *The Grizzlies of Glacier*. Missoula, MT: Mountain Press Publishing. 1988.

Haselhorst, Sarah, and Sarah Claire McDonald. Living With Alligators: After Another SC Attack, How Do We Coexist? What Should Be Done? *The Island Packet*. August 30, 2022.

Hawken, Paul. *Regeneration: Ending the Climate Crisis in One Generation*. New York: Penguin Books. 2021.

Hawkes, Alison. June 19, 2014. https://oaec.org/wp-content/uploads/2020/11/OAEC-Beaver-in-CA-4.0.pdf ps://baynature.org/article/beavers-used-to-be-almost-everywhere-in-california/. (Accessed May 27, 2022).

Heinz, Mark. Bad Predator Policies in Montana, Idaho Could Derail Delisting Wyoming Grizzlies, Bear Experts Say. *Cowboy State Daily*. March 29, 2023. https://cowboystatedaily.com/2023/03/29/bad-predator-policy-in-montana-idaho-could-derail-delisting-wyoming-grizzlies-bear-expert-says/. (Accessed March 30, 2023).

——. Officials Warn People Who Ignore 'Do Not Enter' Grizzly Study Area Signs Can Become Dinner for Bears. *Cowboy State Daily*. May 12, 2023. https://cowboystatedaily.com/2023/05/12/people-who-ignore-do-not-enter-grizzly-study-area-signs-can-become-dinner-for-angry-bears/. (Accessed May 21, 2023).

Helmore, Edward. Summer of Shark: Spate of Encounters Leaves New York Beachgoers Wary. *The Guardian*. July 26, 2022. https://www.theguardian.com/environment/2022/jul/25/shark-encounters-long-island-new-york. (Accessed August 2, 2022).

Herrick, Francis. The Eagle in Action: An Intimate Study of the Eyrie Life of America's National Bird. *National Geographic*. Vol. 15, No. 5. May 1929. 635–660.

Holland, Jennifer S. Wildlife Crossings Can Protect Migrating Animals. *Pew Trust*. https://www.pewtrusts.org/en/trust/archive/spring-2020/wildlife-crossings-can-protect-migrating-animals. June 10, 2020. (Accessed May 15, 2022).

Hornaday, William T. *Extermination of the American Bison*. Reprinted from: Report of the National Museum. 1886-'87. GOVT PRINTING OFFICE. 1889.

——. *The American Natural History*. New York: Charles Scribner's Sons. 1904.

Houck, Oliver A. Astor's Dream: The Great Northwest Fur Emporium. *Environmental Forum*. Vol. 33, No. 5. September–October 2016. 6–7.

Hynes. Thomas. The Boogie Down Beavers of NYC'S Bronx River. *Untapped New York*. May 10, 2021. https://untappedcities.com/2021/05/10/beavers-nyc-bronx-river/. (Accessed May 29, 2022).

HBC History Foundation. Maud Watt. https://www.hbcheritage.ca/people/women/maud-watt. (Accessed May 22, 2022).

——. The Royal Charter. https://www.hbcheritage.ca/things/artifacts/the-royal-charter#:~:text=Granted%20by%20King%20Charles%20II,Company%20of%20Adventurers%20over%20their. (Accessed May 7, 2022).

——. David Thompson. https://www.hbcheritage.ca/people/explorers/david-thompson. (Accessed May 30, 2022).

Jachowski, David. *The Conversation*. April 26, 2023. https://theconversation.com/in-protecting-land-for-wildlife-size-matters-heres-what-it-takes-to-conserve-very-large-areas-201848. (Accessed April 29, 2023).

Jefferson, Thomas. *Writings*. Merrill D. Peterson, ed. New York: Literary Classics of the United States. 1994.

Johnston, Jeremy M. *Introduction in Ernest Thompson Seton's Wahb: The Biography of a Grizzly*. Norman: University of Oklahoma Press. 2015.

Jones, Barbara K. *Bogged Down: The Issue of Wetland Expansion for New Jersey and Massachusetts Cranberry Growers, Environmentalists, and Regulators*. Dissertation. 2000.

——. *Wild Capital: Nature's Economic and Ecological Wealth*. Gainesville: University of Florida Press. 2019.

——. Unifying America One Bison at a Time. *Anthropology News website*. March 26, 2021.

——. Invented Boundaries. *Anthropology News website*. April 11, 2022.

Jordan, Peter A., James L. Nelson, and John Pastor. Progress Towards the Experimental Reintroduction of Woodland Caribou to Minnesota and Adjacent Ontario. The Seventh North American Caribou Conference, Thunder Bay, Ontario, Canada, 19–21 August, 1996. https://septentrio.uit.no/index.php/rangifer/article/view/1554/1459.

Kane, Matt. Two New Positions Hired to Work with Communities and Farmers on Maryland's Eastern Shore. *The Nature Conservancy*. December 2, 2021. https://www.nature.org/en-us/newsroom/md-new-staff-eastern-shore/. (Accessed February 9, 2023).

Kantner, Seth. *A Thousand Trails Home: Living with Caribou*. Seattle, Washington: Mountaineer Books. 2021.

Kellert, Stephen R. *Birthright: People and Nature in the Modern World*. New Haven: Yale University Press. 2012.

Kemp, Martin. *The Human Animal in Western Art and Science*. Chicago: University of Chicago Press, 2007.

Kesselheim, Alan. Yellowstone's Wolf Toll. *Montana Quarterly*. Summer 2022. 16–23.

Keulartz, Josef. Using Metaphors in Restoring Nature. *Nature and Culture*. Vol. 2, No. 1. Spring 2007. 27–48.

Killing Wolves. *Historic Ipswich*. https://historicipswich.org/2021/04/25/killing-wolves/. (Accessed February 9, 2023).

Kirby, Matthew. California to Colorado. *National Parks*. Vol. 97, No. 1. Winter 2023.

Knight, John. *Waiting for Wolves in Japan: An Anthropological Study of People-Wildlife Relations*. Honolulu: University of Hawaii Press. 2006.

Koshmrl, Mike. Yellowstone Grizzly Numbers Jump with Count Revision. *Cody Enterprise*. April 12, 2021. https://www.codyenterprise.com/news/local/article_cb663ed6-9bbc-11eb-b44f-ebfef988c7ec.html. (Accessed March 13, 2023).

——. Wyoming May Pursue Grizzly Delisting Alone. *Jackson Hole News & Guide*. October 6, 2021. 10A.

——. Wyoming lawmakers bring two grizzly bear bills as future of federal protection grows murky. *WyoFile*. January 29, 2025. https://wyofile.com/wyoming-lawmakers-bring-two-grizzly-bear-bills-as-future-of-federal-protection-grows-murky/. (Accessed February 2, 2025).

Krech III, Shepard. *The Ecological Indian: Myth and History*. New York: W.W. Norton & Co. 1999.

Kristof, Nicholas. Make Way for Buffalo. *New York Times Opinion*. Oct 29, 2003. https://www.nytimes.com/2003/10/29/opinion/make-way-for-buffalo.html. (Accessed January 20, 2021).

Lakoff, George, and Mark Johnson. *Metaphors We Live By*. Chicago: The University of Chicago Press. 1980.

Lanman, Christopher, et al. The Historical Range of beaver in the Sierra Nevada: A review of the Evidence. *California Fish and Game* 98 (2): 65–80, 2012.

Lanman, Christopher et al. The Historical Range of Beaver in the Coastal CA: An Updated Review of the Evidence. *California Fish and Game* 99 (4) 193–221, 2012.

Leggett, Doreen. Orleans Creates Shark Safety Video. https://capecod.wickedlocal.com/news/20160814/orleans-helps-create-shark-safety-video. (Accessed March 27, 2019).

Leopold, Aldo. *A Sand County Almanac: With Essays on Conservation from Round River*. New York: Ballantine Books. 1970.

——. The Conservation Ethic. In *The River of the Mother of God and Other Essays by Aldo Leopold*. Susan L. Flader and J. Baird Callicott, eds. Madison, WI: University of Wisconsin. 1933.

Lesak, Stephen. All My Relations: Buffalo, Wild Horses and the Politics of Belonging on the Wind River Reservation. In *High Country News*. Vol. 53, No. 12. December 2021. 28.

Loo, Tina. *States of Nature: Conserving Canada's Wildlife in the Twentieth Century*. Vancouver: University of British Columbia. XI. 2006.

Lopez, Barry. *Of Wolves and Men*. New York: Scribner. 1979.

Lundquist, Kate, and Brock Dolman. Beaver in California: Creating a Culture of Stewardship. *Occidental Arts and Ecology Center Water Institute*. 2020. 4. https://oaec.org/wp-content/uploads/2020/11/OAEC-Beaver-in-CA-4.0.pdf. (Accessed May 27, 2022).

Mallory, F. F., and T. L. Hillis. Demographic Characteristics of Circumpolar Caribou Populations: Ecotypes, Ecological Constraints, Releases, and Population Dynamics. *Rangifer*. Special Issue, No. 10. 1998. 149–160.

Manfredo, Michael J. et al. Social value shift in Favor of Biodiversity Conservation in the United States. *Nature Sustainability*. December 7, 2020. https://rewberl.github.io/public/pdf/manfredo2020social_value_shift_in_favour_of_biodiversity_conservation.pdf. (Accessed April 7, 2022).

Marris. Emma. A Basic Premise of Animal Conservation Looks Shakier Than Ever. *The Atlantic*. February 22, 2023. https://www.theatlantic.com/science/archive/2023/02/wild-animal-species-conservation-native-range-habitats/673153/. (Accessed March 26, 2023).

Matthew 7:15–20.

Mattson, David. Worldviews and the Ethos of State Wildlife Management. In *Grizzly Times*. Vol. 3. 10–19. https://www.grizzlytimes.org/_files/ugd/d2beb3_a4aff4d546d941b08341d6f7d27e66ad.pdf. (Accessed April 20, 2022).

——. The Cult of Hunting And its Timely Demise. In *Grizzly Times*. Vol. 4. 20–28. https://www.grizzlytimes.org/_files/ugd/d2beb3_a4aff4d546d941b08341d6f7d27e66ad.pdf. (Accessed April 20, 2022).

——. Sociopolitical Entanglements of State Wildlife Management. In *Grizzly Times*. Vol. 5. 29–41. https://www.grizzlytimes.org/_files/ugd/d2beb3_a4aff4d546d941b08341d6f7d27e66ad.pdf. (Accessed April 20, 2022).

McCay, Bonnie J., and James M. Acheson. Human Ecology of the Commons. In *The Question of the Commons: The Culture and Ecology of Communal Resources*. Bonnie J. McCay and James M Acheson, eds. Tucson: AZ: The University of Arizona Press. 1996. 1–34.

McCoulough, Mark. An Evaluation of the Maine Caribou Reintroduction Project, 1986–1989. Maine Caribou Project. 1990.

——. Lessons in Reintroduction Policy from an Attempt to Restore Caribou to Maine. Maine Caribou Project. 1991.

McCracken, Harold, and Harry Van Cleve. *Trapping: The Craft and Science of Catching Fur-bearing Animals*. New York: A. S. Barnes & Co. 1947.

McGlashen, Andy. The Fight to Save Lake Superior's Last Caribou. *Sierra*. August 7, 2021. https://www.sierraclub.org/sierra/fight-save-lake-superior-s-last-caribou.

McHugh. Tom. *The Time of the Buffalo*. New York: Knopf. 1979.

McKean, Andrew. Elk Inside the Frame. *Bugle*. March–April 2023. 58–66.

McNally, Bob. Maine is Asking Whitetail Hunters to Kill More Does. https://outdoorlife. com. October 8, 2021. (Accessed March 15, 2022).

Meine, Curt. A Letter to the Editors: In Defense of the Relative Wild. In *After Preservation: Saving American Nature in the Age of Humans*. Ben A. Minteer and Stephen J. Pyne, eds. Chicago: University of Chicago Press. 2015. 84–93.

Methow Beaver Project Accomplishments and Outcomes. April 2013. http://www.methowrestorationcouncil.org/MethowBeaverProjectReport2013.pdf. (Accessed September 2, 2021).

Mitsch, William J., and James G. Gosselink. *Wetlands*. New York: Van Nostrand Reinhold. 1993.

Mohr, Kylie. Get to Know the Whitebark Pine. *High Country News*. March 2023. 9.

Money Substitutes in New Netherland and Early New York: The Beaver Pelt. https://coins.nd.edu/colcoin/colcoinintros/nnbeaver.html. (Accessed May 13, 2022).

Montana Grizzly Bear Draft Management Plan 2022. https://fwp.mt.gov/binaries/content/assets/fwp/aboutfwp/public-comments/grizzly-bear-management/statewide-grizzly-bear-management-plan-12.6.22.pdf. (Accessed July 10, 2024).

Moore, C. C. *A Visit from St. Nicholas*. United Kingdom: Henry M. Onderdonk. 1848.

Morgan, Lewis H. *The American Beaver and His Works*. London: Forgotten Books. 1868.

Moser, Karin S. Metaphor Analysis in Psychology- Method, Theory, and Fields of Application. *Forum: Qualitative Social Research*. Vol. 1, No. 2. Article 21. June 2000.

Nash, Roderick Frazier. *Wilderness & the American Mind*. New Haven: Yale University Press. 1982.

NRDC. Canadian Logging Takes Alarming Climate Toll. In *Nature's Voice*. Spring 2023.

National Park Service. Draft Environmental Impact Statement for Bison Management. https://parkplanning.nps.gov/document.cfm?parkID=111&projectID=94496&documentID=130873. August 10, 2023. (Accessed October 10, 2023).

National Wildlife Refuge Association. http://refugeassociation.org/about/about-the-refuge-system/. (Accessed January 22, 2021).

NCASI. 2020. Current State of Knowledge and Research on Woodland Caribou in Canada. Technical Bulletin No. 1066. Cary, NC: National Council for Air and Stream Improvement, Inc.

Niemi, Ernie, and Suzanne Fouty. Economic Benefits of Beaver- Created and Maintained Habitat and Resulting Ecosystem Services. *Executive Summary*. August 25, 2021. Unpublished.

NOAA. *Risk Communication Basics*. Social Science of Coastal Programs, 2016.

Norris, P. W. Annual Superintendent Report Yellowstone National Park. 1880.

Northwest Territories Species at Risk Committee 5. https://www.nwtspeciesatrisk.ca/sites/enr-species-at-risk/files/polar_bear_status_and_reassessment_report_final_april2021.pdf. (Accessed April 22, 2022).

Olwig, Kenneth R. Reinventing Common Nature: Yosemite and Mt. Rushmore—a Meandering Tale of a Double Nature. Paper Presented at the Fifth Annual Common Property Conference of the International Association for the Study of Common Property. Bodo, Norway. May 24–28, 1995. Session on Rethinking the Commons.

O'Neill, Peggy. Leaving Some for the Fish. *Montana Outdoors*. May–June 2022. 31–35.

Organ, J. F., et al. *The North American Model of Wildlife Conservation*. The Wildlife Society and The Boone and Crockett Club Technical Review 12–04–December 2012. https://wildlife.org/wp-content/uploads/2014/05/North-American-model-of-Wildlife-Conservation.pdf.

Ostrom, Elinor. *Governing the Commons: The Evolution of Institutions for Collective Action*. Cambridge: Cambridge University Press. 1990.

Pagnamenta, Peter. *Prairie Fever: British Aristocrats in the American West 1830–1890.* New York: W.W. Norton & Company. 2012.

Park County Environmental Council. 2022 Wolf Trapping and Hunting Quota Talking Points.

Partnow, Patricia H. *Counting on Caribou*. Guilford, CT: Muddy Boots. 2020.

Pashby, Jim. Stymied: A Grizzly's Journey Highlights the Need for More Wildlife Road Crossings. *Montana Outdoors*. March–April 2022. 20–21.

Perryman, Heidi. https://humansandnature.org/heidi-perryman/. (Accessed June 5, 2022).

Personal Memoirs Ulysses S. Grant. https://www-tc.pbs.org/wgbh/americanexperience/media/filer_public/00/9d/009d03b6-2f0c-4197-a13d-219044df1182/grant_memoirs.pdf. (Accessed April 13, 2022).

Peterson, Christine. Elkonomics of Elk Antlers. *The Bugle*. September–October 2021. 37–41.

——. Chronic Mystery. *High Country News*. April 2023. 34–40.

Phillips, Taylor. WYIdlife Tourism for Tomorrow. https://www.wildlifetourismfortomorrow.org/blog/wildlife-tourism-for-tomorrow-initiative-launches-to-support-conservation. (Accessed May 12, 2022).

Pinchot, Gifford. *Breaking New Ground.* New York: Harcourt Brace. 1947.

Pinelands Adventures. https://www.pinelandsadventure.org. (Accessed May 29, 2022).

Preston, Christopher J. *Tenacious Beasts: Wildlife Recoveries That Change How We Think about Animals*. Cambridge, MA: MIT Press. 2023.

PBS Nature. Wolf Wars: America's Campaign to Eradicate the Wolf. September 14, 2008. https://www.pbs.org/wnet/nature/the-wolf-that-changed-america-wolf-wars-americas-campaign-to-eradicate-the-wolf/4312/. (Accessed April 2, 2022).

Recovery Strategy for the Grizzly Bear (Ursus arctos), Prairie Population, in Canada. Environment Canada. 2009.

Repanshek, Kurt. Michigan Man Fatally Mauled by Grizzly in Yellowstone National Park Tried to Defend Himself. March 5, 2012. https://www.nationalparkstraveler.org/2012/03/michigan-man-fatally-mauled-grizzly-yellowstone-national-park-tried-defend-himself957. (Accessed April 10, 2022).

Roach, Mary. *Fuzz: When Nature Breaks the Law*. New York: W.W. Norton & Co. 2021.

Robbins, Jim. Mass Yellowstone Hunt Kills 1,150 Bison. *New York Times*. April 4, 2023. https://www.nytimes.com/2023/04/04/science/bison-hunt-yellowstone-native-americans.html.

Rosner, Mary, and T. R. Johnson. Telling Stories: Metaphors of the Human Genome Project. *Hypatia*. Vol. 10, No. 4. Autumn 1995. 104–129.

Rovelli, Carlo. The Best Reason to Trust Science. *Wall Street Journal*. March 11–12, 2023. C5.

Ruiz, Santi. Bring Back the Bison. *National Review*. September 13, 2020. https://www.nationalreview.com/2020/09/bring-back-the-bison/. (Accessed January 25, 2021).

Rump, Joseph. Wolf Personalities: Ecosystem Impacts and the Possibility of Wolf Cultures. Voyageurs Wolf Project 2023 Winter Webinar Series. March 15, 2023.

Rural Montana Report. Improving Elk Management in Montana. February 2021. http://www.ruralmt.org/wp-content/uploads/Improving-Elk-Management-in-Montana.pdf. (Accessed March 2, 2022).

Said, Edward W. *Orientalism*. New York: Vintage Books. 1979.

Save the Nautilus Film. Save the Nautilus Conservation Organization. New York City Wildlife Conservation Film Festival. 2021.

Schwartz, Judith D. *The Reindeer Chronicles and Other Inspiring Stories of Working with Nature to Heal the Earth*. White River Junction, VT: Chelsea Green Publishing. 2020.

Sentinel: American Prairie. A Conversation on Conservation with Kris Tompkins. Summer 2023. 4–9.

Seton, Ernest Thompson. *The Arctic Prairies*. BiblioBazaar. Reprint/Original publication 1911.

Sillars, Jordan. The Blueprint for Banning Beaver Trapping. *MeatEater*. The Blueprint for Banning Beaver Trapping | MeatEater Conservation (themeateater.com) September 24, 2021. (Accessed September 8, 2023).

Soniak, Matt. The Origins of Our Misguided Hatred for Pigeons. *Audubon*. November 14, 2016. https://www.audubon.org/news/the-origins-our-misguided-hatred-pigeons.

Stamm, Keith R., and James E. Grunig. Communication Situations and Cognitive Strategies in Resolving Environmental Issues. In *Journalism and Mass Communication Quarterly*. December 1977. file:///C:/Users/docjo/Downloads/CommunicationSituationsandCognitiveStrategiesinResolvingEnvironmentalIssues JMCQ1977.pdf. (Accessed April 25, 2021).

Stern, Richard O., et al. Eskimos, Reindeer, and Land. School of Agriculture & Land Resources Management. Bulletin 59. December 1980.

Stinson, Ann. *The Ground at My Feet: Sustaining a Family and a Forest*. Corvallis: Oregon State University Press. 2021.

Stinson, D. W., and G. J. Wiles. Draft Periodic Status Review for the Woodland Caribou in Washington. Olympia, Washington: Washington Department of Fish and Wildlife. 2022. 23+iii.

Survey Finds Growing Tolerance Among Montanans or Wolves. Montana Fish, Wildlife, and Parks and University of Montana's College of Forestry and Conservation. January 11, 2024. https://www.umt.edu/news/2024/01/011124wolf.php. (Accessed May 4, 2024).

Suzuki, David. *Billion Dollar Caribou*. aired March 21, 2013. S52, E13.

Taraldsen, Lars Erik, Paulsson, Lars, and Jesper Starn. Reindeer Herders Push to Reclaim Land from Norway Wind Farms. *Bloomberg*. February 21, 2022. https://www.bloomberg.com/news/articles/2022-02-21/reindeer-herders-push-to-remove-norway-wind-turbines.

The Economist/YouGov (2018). March 10-13, 2018, poll of 1500 US adults. Topic 48J. Moral Acceptability of Various Behaviors - Hunting Animals for Sport.

The Mountain Caribou Technical Advisory Committee. *A Strategy for the Recovery of Mountain Caribou in British Columbia*. British Columbia Ministry of Water, Land and Air Protection. September 2002. https://www.env.gov.bc.ca/wld/documents/mtcaribou_rcvrystrat02.pdf. (Accessed December 27, 2022).

The Nature Conservancy. Post–Sandy: Natural Habitats Vital for Protecting our Coasts. https://www.nature.org/en-us/about-us/where-we-work/united-states/new-jersey/stories-in-new-jersey/sandy-one-year-later/. (Accessed January 2, 2022).

THE WILDERNESS ACT Public Law 88-577 (16 U.S.C. 1131-1136) 88th Congress, Second Session September 3, 1964. https://winapps.umt.edu/winapps/media2/wilderness/NWPS/documents/publiclaws/The_Wilderness_Act.pdf. (Accessed September 22, 2022).

Thomas, Heather S. Producer Partnership Provides Meat for Needy People. *Farm Progress*. February 3, 2022. https://www.farmprogress.com/livestock/producer-partnership-provides-meat-for-needy-people. (Accessed November 20, 2024).

Thompson, Paul B. *The Spirit of the Soil: Agriculture and Environmental Ethics*. New York: Routledge. 1995.

Thoreau, Henry D. A Week in the Concord and Merrimack Rivers. In *The Nature Writings of Henry David Thoreau*. Ann Harbor: Tally Hall Press. 1996. 49.

Tilley, Christopher. *Metaphor and Material Culture*. Oxford: Blackwell Publishers. 1999.

Tong, Ziya. The Amazing Sea Wolves of the Great Bear Rainforesmt. *Best of Canadian Geographic*. Special Collector's Edition. July/August 2021. 30–37.

Trippensee, Reuben E. *Wildlife Management: Upland Game and General Principles*. Volume 1. New York: McGraw-Hill. 1948.

USDA APHIS. https://www.aphis.usda.gov/aphis/ourfocus/wildlifedamage/operational-activities/sa_beavers/ct_conflicts. (Accessed June 2, 2022).

van Zelst, Britt. Beavers in Anthropogenic Landscapes: A Blessing or a Burden? unpublished graduate scientific proposal. July 2022. 1.

Wacker, Peter O., and Paul G. E. Clemens. *Land Use in Early New Jersey: A Historical Geography*. Newark: New Jersey Historical Society. 1995.

Western Arctic Caribou Herd Working Group. *Caribou Trails*. Vol 15. Summer 2015. https://westernarcticcaribou.net/wp-content/uploads/2016/09/caribou-trails-2015.pdf.

Wildlife Museum. What are the Odds of a Shark Attack? http://dev.thewildlifemuseum.org/exhibits/sharks/odds-of-a-shark-attack/. (Accessed April 16, 2019).

Wild Livelihoods. 2022 Tourism Study, Park County. RRC Associates and ITRR. Park County Tourism Study | Wild Livelihoods Business Coalition. (Accessed August 10, 2022).

Wilkinson, Todd. Trying to Balance Wolves. *Big Sky Journal*. Fall 2015.

——. To Live or Die in Bear Country: Counting the Seconds in Your Grizzly Moment of Truth. *Mountain Journal*. October 29, 2017. https://mountainjournal.org/to-live-and-die-in-grizzly-country. (Accessed September 22, 2021).

——. The Killing Fields Await Yellowstone Bison Once Again. *Mountain Journal*. December 15, 2017. https://mountainjournal.org/the-killing-fields-return-for-yellowstone-bison. (Accessed May 15, 2021).

Wolf Management Plan. Wind River Reservation. Eastern Shoshone and Northern Arapaho Tribes Ft. Washakie and Ethete, WY Shoshone and Arapaho Tribal Fish and Game Department Ethete, WY Assisted by the U.S. Fish and Wildlife Service Management Assistance Office Lander, WY April 13, 2007.

Wolf Wars: America's Campaign to Eradicate the Wolf. *Nature*. September 14, 2008. https://www.pbs.org/wnet/nature/the-wolf-that-changed-america-wolf-wars-americas-campaign-to-eradicate-the-wolf/4312/. (Accessed March 4, 2022).

Wolfe, Linnie Marsh. *Son of the Wilderness: The Life of John Muir*. Madison: University of Wisconsin Press. 1978.

Woodard, Colin. *American Nations: A History of the Eleven Rival Regional Cultures of North America*. New York: Viking Press. 2011.

——. The Frontier That Wasn't. *Smithsonian*. Vol. 53, No. 7. January–February 2023. 20–24.

Worlddeer.org. Caribou vs. Reindeer Comparison. https://worlddeer.org/caribou-vs-reindeer/.

Wyoming Game and Fish Department News. April 15, 2024. https://wgfd.wyo.gov/news-events/grizzly-bear-euthanized-due-cattle-depredation-washakie-county. (Accessed April 18, 2024).

Yellowstone Bison: Conserving an American Icon in Modern Society. Edited by P. J. White, Rick L. Wallen, and David E. Hallac. The Yellowstone Association, Yellowstone National Park. 2015. https://www.nps.gov/yell/learn/nature/upload/Yellowstone_Bison_ForWeb.pdf. (Accessed April 20, 2022).

Yellowstone Bison: Restoring a North American Icon. Park County Community Conservation Series. September 20, 2023.

2022 Draft MT-FWP Grizzly Bear Management Plan. https://fwp.mt.gov/binaries/content/assets/fwp/aboutfwp/public-comments/grizzly-bear-management/statewide-grizzly-bear-management-plan-12.6.22.pdf. 1–217. (Accessed March 13, 2023).

Index

Note: Page references for figures and tables are italicized

About the Author

Barbara K. Jones has made her career out of both teaching anthropology and researching human perceptions of nature. After earning a bachelor's degree in agricultural economics from Virginia Tech and then a PhD in anthropology from Rutgers University, she has written articles for *Culture, Agriculture, Food and Environment*; *Journal of Ecological Anthropology*; *Anthropological Perspectives on Economic Development and Integration*; *Practicing Anthropology*; and *Anthropology News*. She is also the author of *Wild Capital: Nature's Economic and Ecological Wealth*. Her current research focuses on how by reimagining the metaphors and stories commonly used to describe nature's "inconvenient species," coexistence becomes more likely. She and her husband, Tommy, split their time between New Jersey and Montana.

www.ingramcontent.com/pod-product-compliance
Lightning Source LLC
LaVergne TN
LVHW010719210926
848485LV00002B/78
* 9 7 8 1 6 6 6 9 1 4 8 0 1 *